The Role of Chromosomal Change
in Plant Evolution

Oxford Series in Ecology and Evolution
Edited by Paul H. Harvey and Robert M. May

The Role of Chromosomal Change
in Plant Evolution

Donald A. Levin

UNIVERSITY PRESS

2002

OXFORD
UNIVERSITY PRESS

Oxford   New York
Auckland   Bangkok   Buenos Aires   Cape Town   Chennai
Dar es Salaam   Delhi   Hong Kong   Istanbul   Karachi   Kolkata
Kuala Lumpur   Madrid   Melbourne   Mexico City   Mumbai   Nairobi
São Paulo   Shanghai   Singapore   Taipei   Tokyo   Toronto

Published by Oxford University Press, Inc.
198 Madison Avenue, New York, New York 10016

www.oup.com

Oxford is a registered trademark of Oxford University Press.

Library of Congress Cataloging-in-Publication Data
Levin, Donald A.
   The role of chromosomal change in plant evolution / Donald A. Levin
      p. cm.—(Oxford series in ecology and evolution)
   Includes bibliographical references (p.).
   ISBN 978-0-19-513860-3
   1. Plant chromosomes. 2. Plants—Evolution. I. Title. II. Series.
   QK725 .L464 2002
   581.3′8—dc21   2002190385

Printed in the United States of America
on acid free paper

Preface

Chromosomal change is a prime stimulus for flowering plant evolution. Such change may directly affect the biology of organisms or set the stage for population divergence by establishing barriers to gene flow. These barriers most often take the form of hybrid sterility. Therefore, the more we know about chromosomal change, the better we can understand the process of evolution and its products.

New molecular technologies have greatly increased our ability to document chromosomal change in plants and to understand the direction of this change. For example, there is a broad database on genome size variation within and among species. There is a diverse array of nuclear and cytoplasmic genetic markers to help identify the genomic compositions of polyploids. New chromosomal staining techniques allow us to better recognize karyotypic differences between populations and species, and to better analyze chromosome pairing behavior in hybrids. Comparative genomic mapping has opened a new field of molecular cytogenetics and allows us to document fine-scale chromosomal rearrangements. Using genome sequencing, we can better understand the genetic consequences of obligate chromosomal heterozygosity.

The literature on chromosomal change in wild plants and in domesticates is growing rapidly. Unfortunately, this literature is fragmented, in the sense that the literature on one topic is disconnected from that on other topics, even though they may be related. Although numerous excellent reviews have been written on a broad range of subjects in the 1990s, there is no centralized source of information on chromosomal evolution in plants. There is no place for students or professionals to go for a contemporary synthesis. The last synthesis was by G. Ledyard Stebbins in his 1971 book *Chromosomal Evolution in Higher Plants.*

The purpose of this book is to review and integrate the large body of information on chromosomal evolution in angiosperms and to place this product within an evolution–speciation framework. The book is devoted to exploring the mechanisms promoting chromosomal change and the consequences of these changes. The focal population systems are the species and the genus.

Chapter 1 deals with genome size differences within and among congeneric species. The bases for differences, ecological correlates with genome size differences, and evolutionary trends are the focal points. Chapter 2 treats chromosomal rearrangements within species. The focus is on the types of chromosomal variation, their meiotic signatures, and their participation in chromosome races. Chapter 3 deals with patterns of chromosome number and arrangement changes among diploid congeneric species. Molecular phylogenies allow one to interpret the direction of these changes. Chapter 4 delves into the impact of chromosomal rearrangements and change in chromosome number on the potential for gene exchange between populations and discusses the incorporation of alien chromosome segments. Chapter 5 explores the extent of permanent chromosomal heterozygosity, its genetic correlates, and the mechanisms responsible for its perpetuation, as well as the pathways to permanent heterozygosity. Chapter 6 deals with the occurrence of polyploidy, the types of polyploids, and their modes of establishment. Regarding the latter, special attention is given to the formation of unreduced gametes, the genesis and sterility of triploids, and the problems of reproduction faced by newly emergent and fertile polyploids. Chapter 7 delves into the myriad phenotypic effects of chromosome doubling. Chapter 8 considers how the genetic compositions of autopolyploid populations differ from those of their diploid progenitors and reviews the meiotic problems imposed by chromosome doubling. Chapter 9 explores the evolution of polyploid lineages, as inferred from the numerous molecular tools now available to us. In addition to documenting the genomic compositions of polyploids, chapter 9 explores the evidence that newly emergent polyploids may undergo genetic and chromosomal changes before they become stable entities.

I am very grateful to the assistance given to me during the formulation of this book. I give special thanks to Gábor Lendvai, who read the entire manuscript and provided many insights and useful suggestions. Many thanks to Gerald Carr for reading the chapters on chromosomal rearrangements and to David Herrin for reading the chapter on genome size. Thanks to the series editors, Robert May and Paul Harvey, for their efforts on my behalf. I bear full responsibility for the errors and omissions in the final product.

I also am very grateful to the people who helped in the production of this book. Special thanks to Gerald Carr for providing photographs of *Calycadenia* and *Dubautia* chromosomes, and to Kuniaki Watanabe for providing photographs of the *Brachyscome* and *Podolepis* chromosomes. A huge thanks to

Scott Schulz, who redrew all of the figures used in the book from their original sources. I thank Patricia Watson for her copyediting. Also thanks to Bob Nagy, who kept my computer functioning during the term of the project.

Thanks also to the many copyright holders for their permission to redraw figures.

Finally, I dedicate this book to my Ph.D. mentor, Dale M. Smith.

Austin, Texas   *D.A.L.*
October, 2001

Contents

The Role of Chromosomal Change
in Plant Evolution

1

Heterogeneity in Genome Size

The unreplicated haploid nuclear genome referred to as its C value has been determined in a broad spectrum of angiosperms that encompasses large differences in taxonomic affinity and biological properties. This C value varies over a very broad range, from about 0.2 pg in *Arabidopsis thaliana* to 127.4 pg in *Fritillaria assyriaca* (Bennett and Leitch, 1997). The distribution of genome size in over 2,800 species, as presented by Leitch et al. (1998), is depicted in figure 1.1. Rather than being broadly distributed along the C-value spectrum, most species have less than 10 pg of DNA.

Whether we consider closely related species or remotely related species, genome size does not correlate with the biological complexity of organisms or with their gene number. Nor does genome size vary in concert with evolutionary advancement or with life span or growth form. The independence of genomic size from these variables is referred to as the C value paradox (Thomas, 1993). It is still a subject of considerable interest and debate.

Variation in genome size is accompanied by variation in the length of introns, which are noncoding (intervening) sequences within plant genes. Intron length varies widely among angiosperm species. Might the amount of noncoding DNA residing within genes be related to that residing outside of genes? Vinogradov (1999) showed that there is a compelling correlation between intron size and genome size among major groups of organisms ($r = 0.80$) and that significant correlations are present within groups including angiosperms (figure 1.2). Across groups, the average intron size increases as the genome size increases, with a slope of one-fourth for log-transformed values. What the correlation means is a matter conjecture. Vinogradov suggested that intron size may play a

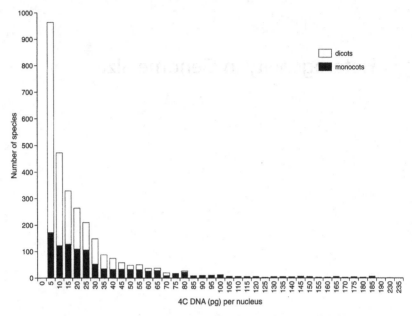

Figure 1.1.   The number of species with different $4C$ values per diploid genome (based on Leitch et al., 1998).

role in the regulation of gene activity through its association with chromatin condensation.

All plants require about the same number of genes (roughly 10^7 to 10^8 bp) to regulate growth, development, and reproduction (Flavell, 1980). However, the number of base pairs per plant genome varies from about 5×10^7 to more than 8×10^{10} (Bennett and Smith, 1976). The divergence of plants in DNA content may have profound implications for a wide range of plant properties and may contribute to their reproductive isolation, as developed below and in other chapters.

Large differences in the genome sizes of related species do not mandate concomitant differences in the number of structural genes. Recombinant mapping shows the conservation of Mendelian genes against a background of variable DNA content in several grasses (Ahn et al., 1993; Moore et al., 1995). For example, diploid barley and rice, which have similar levels of morphological and physiological complexities, differ about 11-fold in genome size but have roughly the same number of genes. Roose and Gottlieb (1978) determined the number of structural genes coding for several electrophoretically detectable enzymes in seven diploid species of *Crepis,* whose genome size varied sevenfold. The number of genes coding these enzymes was about 19, genome size notwithstanding. Only when we compare diploids and polyploids may we find differences in gene number. For example, tetraploid *Tragopogon* has duplicate genes (Roose and Gottlieb, 1976), and hexaploid wheat has triplicate genes (Hart, 1983).

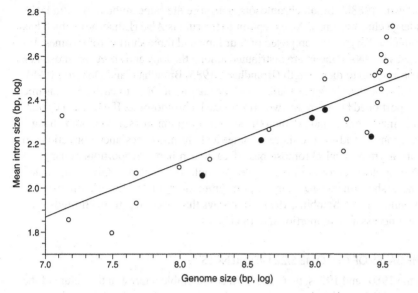

Figure 1.2.   The relationship between the log of genome size and the mean intron size of 27 species. Solid circles denote plant species; open circles, other species. Redrawn from Vinogradov (1999), with permission of Springer-Verlag.

We knew of the vast differences in genome size long before direct measurements became possible. Before the 1960s, genome size comparisons were based on total length of the entire chromosome set. The use of chromosomal data as a proxy for genome size later was validated when the relationship between the length of the chromosome complement could be related to DNA content. The correlations between the two variables typically exceeded $r = 0.85$ within species (e.g., *Vicia,* Raina and Rees, 1983; Ceccarelli et al., 1995), between congeneric species (e.g., *Crepis,* Jones and Brown, 1976; *Lathyrus,* Narayan and Rees, 1976), and among species in related genera (e.g., *Microseris* and *Agoseris,* Price and Bachmann, 1975).

Chromosome size varies dramatically among angiosperms. In a survey of 856 angiosperm species, Levin and Funderberg (1979) found that the mean chromosome length per species ranged from 0.6 μm to more than 14.6 μm, whereas total chromosome length per diploid genome varied from 14.6 μm to more than 250 μm. Given the wide variation in total chromosome length, it is not surprising that species differ in nuclear DNA content.

Joint studies of chromosome complements and DNA contents have helped our understanding of how increases and decreases in genome size are distributed among chromosomes. It is noteworthy that changes in nuclear DNA amounts within genera typically are achieved by equal increments to each chromosome (*Vicia,* Raina and Rees, 1983; *Festuca* and *Lolium,* Seal and Rees, 1982; *Papaver,* Srivastava and Lavania, 1991; *Lathyrus,* Narayan and Durrant, 1983;

Narayan, 1988b). Small chromosomes acquire the same amount of extra DNA as large chromosomes. An exception to the rule is *Aloe* (Liliaceae), which has a bimodal karyotype composed of four long and three short chromosomes. Increases in DNA content are distributed among the long and short chromosomes in proportion to their length (Brandham, 1983; Brandham and Doherty, 1998).

It is not clear how the addition or subtraction of DNA to each chromosome is apportioned between the two arms of each chromosome. If each arm had the same incremental change, then the symmetry of chromosomes would change. Additions would lead to more symmetrical chromosomes and subtractions to more asymmetrical chromosomes. If each arm had a proportional change in DNA content, chromosome symmetry would be unaltered. Given that many genera show increasing karyotype asymmetry with declining genome size (Avdulov, 1931; Stebbins, 1971), it follows that DNA loss from chromosomes is not necessarily proportional to their sizes.

THE BASES OF GENOME SIZE DIFFERENCES

In the 1980s and 1990s, there has been considerable interest in the nature of the nuclear DNA that is above and beyond that necessary for a plant's basic function. This DNA is nongenic, but whether it is functional is a matter of conjecture, as discussed below. The major fraction of the nuclear genomes of most species consists of repetitive DNA elements (Flavell, 1986; Kubis et al., 1998; Schmidt and Heslop-Harrison, 1998). These elements are composed of about 30 distinctive sequence motifs that range in size from 1 bp to more than 10,000. A given motif may represent from 5% to as much as 50% of a genome, being repeated from several hundred to several hundred thousand times. Large blocks of repetitive DNA appear to contribute little to recombination-based genetic distances between markers. As shown in maize, single-copy genes behave in general as hot spots for recombination (Civardi et al., 1994; Dooner and Martinez-Ferez, 1997).

In general, there is a positive correlation between genome size and the percentage of repetitive DNA (Flavell, 1986; Lapitan, 1992; Bennetzen and Kellogg, 1997). *Arabidopsis* has about 14% of its genome in the form of repeated DNA sequences, whereas wheat, whose genome is roughly 600 times larger, has as much as 80% of its genome as repeated sequences.

Repetitive elements can be divided (for the most part) into two groups that are distinguished by their organization and position on chromosomes (Kubis et al., 1998). In one group, individual copies are organized in tandem repeats. The second group of repetitive DNA sequences are dispersed throughout the genome and sometimes are interspersed with other sequences.

Tandem repeats

Tandem repeats include rDNA, different types of satellite DNA, and the telomeric repeat at the ends of chromosomes. These tandemly repeated entities are located primarily in the pericentric, telomeric, subtelomeric, or intercalary regions of chromosomes and are present on all or most chromosomes in the genome. They are likely to have arisen by unequal recombination events or replication slippage (SanMiguel and Bennetzen, 1998). Many tandem repeats have a complex structure. Tandem arrays of repeats constitute the material of heterochromatin (Flavell, 1980).

Genes coding for the 18S, 5.8S, and 25S ribosomal RNAs are one of the major families of repeated sequences in plants (Lapitan, 1992). The ribosomal genes are organized at a small number of sites in the genome. Each repeating unit consists of a transcribed element and an intergene spacer. Lengths for the rDNA repeating unit range from 7.8 kb to 18.5 kb, with copy numbers from 600 to 8,500 per haploid genome (Lapitan, 1992). However, not all copies are transcriptionally active. In *Arabidopsis thaliana,* the rDNA with a basic repeat unit of 10 kb occurs several hundred times and constitutes about 8% of the genome (Pruitt and Meyerowitz, 1986). The sequence of the transcribed element is conserved across the plant kingdom, whereas the interspacer DNA is highly variable.

In the sugar beet, the clusters of 18S-5.8S-25S rRNA genes and intergenic spacers are confined to the secondary constriction at the end of the short arm of chromosome 1 (Schmidt et al., 1994). These genes are associated with the nucleolar-organizing region(s) of the chromosome.

The 5S ribosomal RNA genes are another highly conserved family of repeated sequences in plants (Lapitan, 1992). The 5S DNA occurs separately from the other ribosomal genes and is not associated with the nucleolar-organizing regions of chromosomes. The coding region of the 5S gene is about 120 bp long and, like other rDNA genes, is highly conserved, whereas the intergene spacer varies substantially in length and sequence among different species.

Telomeric DNA consists of conserved 7-bp repeats (TTTAGGG) in most plants (Kubis et al., 1998). These terminal entities are not replicated from pre-existing DNA by semiconservative replication as are other nuclear DNA sequences, but rather are appended to the ends of chromosomes by the enzyme telomerase. Telomeric DNA usually extends for a few kilobases. In many species, the telomeric repeats are associated with another tandemly repeated family (Lapitan, 1992). Telomeric DNA has unique structural features that stabilize chromosomes and allow them to replicate completely without the loss of terminal nucleotides.

Satellite DNA families are groups of identical or similar sequences that are organized into tandemly repeating blocks (Hemleben, 1990). A genome may contain many different satellite DNA families. The lengths of the monomeric unit of these families tend to cluster around 150–180 bp and 320–360 bp. Satellite DNA families can show species and even chromosome specificity, which makes them useful probes for phylogenetic studies. Nine different families have been isolated from different *Beta* species (Kubis et al., 1998). In *Arabidopsis,* the satellite DNA has 6,000 copies of a 185-bp unit, and in tomato it has 75,000 copies of a 162-bp unit (Lapitan, 1992). This DNA is often in constitutive heterochromatin, which usually is found close to the telomeres and centromeres. Centromeric repeats with arrays of 140–360 bp monomers often totaling more than 1,000,000 bp have been reported in *Arabidopsis thaliana* (Murata et al., 1994) and *Beta vulgaris* (Schmidt and Metzlaff, 1991).

Dispersed repeats

A given sequence may occur throughout the genome (e.g., TGRII in tomato) or only at a few chromosomal regions (e.g., TGRIII in tomato; Ganal et al., 1988). The rates of interspersion vary across species. Typically, species with relatively large genomes (e.g., maize and rye) have a short interspersion pattern characterized by short repetitive sequences (5–2,000 bp) interspersed with short single copy DNA (200–4,000 bp; Lapitan, 1992). Conversely, species with small genomes exhibit a long period interspersion pattern, where the single copy sequence length may exceed 120,000 bp.

Interspersed sequences also include mobile DNA elements (transposable elements and retroelements). DNA transposable elements migrate to new locations via DNA intermediates, whereas retroelements use RNA intermediates for transposition. Retroelements are the most abundant class of dispersed repeats. They can be divided into three subgroups: retrotransposons, retroposons, and retrosequences. The prime difference between retrotransposons and retroposons is the presence or absence, respectively, of long terminal repeats (LTRs). In contrast to these subgroups, retrosequences cannot move autonomously. Several full-length retrotransposons have been described in plants, and they range in size from 4.8 kb in maize to 15.5 kb in sorghum (Bennetzen, 1996). Retrotransposons account for between 49% and 78% of the nuclear DNA in maize (SanMiguel and Bennetzen, 1998). Five different retrotransposon families comprise over 25% of the maize genome (Bennetzen and Kellogg, 1997).

Any single retroelement appears to be limited to a narrow range of species. When a given element occurs throughout a genus, the copy number may vary dramatically between species (Bennetzen, 1996). Retroelements may insert in a gene and sometimes cause its inactivation. Insertion near genes may affect their expression or recombination rates. In maize, most retrotransposon insertions appear to be into older retrotransposons (Walbot and Petrov, 2001). This pattern may be due to the avoidance of genes as insertion sites or because the nonfunctional genes caused by the insertion are eliminated by natural selection.

The second major class of dispersed repeats are transposable elements that replicate through a DNA intermediate. These elements typically have copy numbers that range between 1 and 100. Examples include *Ac, Spm,* and *Mu* in maize (Döring and Salinger, 1986), *Tph1* in petunia (Gerats et al., 1990), and *Tag1* in *Arabidopsis thaliana* (Tsay et al., 1993).

Transposable elements ostensibly are an important source of spontaneous mutations (Kunze et al., 1997). They may induce deletions, insertions, frameshifts, inversions, translocations, and duplications. Their activity is enhanced by "genomic stress," which might be forthcoming from inbreeding or hybridization (McClintock, 1984). These elements thus may have an important role as evolutionary catalysts (McDonald, 1995).

PHENOTYPIC CONSEQUENCES OF GENOME SIZE DIFFERENCES

Given that species may differ manifestly in their genome sizes and that repetitive DNA may account for much or all of the difference, there has been considerable interest in the effect that repetitive DNA has on the organism. There are two opposing viewpoints on this matter. One viewpoint, elaborated by Orgel and Crick (1980) and Doolittle and Sapienza (1980), is that noncoding DNA essentially is "junk" DNA. Its only function is self-preservation and growth, somewhat like a parasite. Its alteration has a negligible effect on the organism. Repetitive DNA levels, and thus genome size, are driven by factors such as gene amplification, gene conversion, unequal crossing over, and transposition and will tend to increase until the costs of replicating it become too great (Walsh, 1985; Charlesworth, 1987). At that point, a balance will be struck between the mutational forces that increase genome size and the selective forces acting on the organism to reduce genome size. The weaker the selective forces against junk DNA, the larger the predicted genome size, all else being equal.

A second view of genome size is that repetitive DNA has a function and thereby an adaptive role in various aspects of plant structure and function. Nuclear DNA thus may influence the phenotype through the expression of its genic content and also by the physical effects of its volume and mass. The term "nucleotype" defines those properties of DNA that affect the phenotype independently of its encoded (low-copy DNA) information content (Bennett, 1972).

Nucleotypic effects of DNA content variation were first indicated by surveys of unrelated species. Beginning at the cellular level, there is a striking positive correlation between DNA C values and the volumes of the nucleus and nucleolus, cell size and mass, pollen volume, and chloroplast number per guard cell (Bennett, 1987). Genome size variation also is positively correlated with mitotic and meiotic cycle times and rate of pollen development. The longer cell cycle time in species with larger genome sizes usually is ascribed to the time needed to replicate a larger amount of DNA.

Because cell size and minimum cell doubling time interact to determine the rate of growth (or development), the minimum time from germination to the production of the first seed (i.e., the minimum generation time) is the least in species with the lowest DNA content. These are the ephemeral annuals, whose minimum generation time is less than 7 weeks. As we go to longer lived annuals, then facultative perennials, and finally obligate perennials, DNA content tends to increase (Bennett, 1972).

Genome size also correlates with phenological expression. Grime and Mowforth (1982) found that the earlier the time of shoot expansion (February to July) in the flora of Sheffield, England, the greater the DNA content. They also found that mean leaf extension per day positively correlated with DNA content. Thus, growth of plants in the cold spring weather is superior in species with higher DNA content. A similar relationship was noted in the North Derbyshire (England) flora (Grime et al., 1985).

Seed size and germination temperature also are related to genome size. There is a tendency for large seeds to have relatively high DNA contents and to germinate at lower temperatures (Thompson, 1990). The latter occurs in part because germination and early growth both take place largely by cell expansion.

The nucleotypic effects of genome size are evident in crops as well as in wild plants. Consider, for example, a series of *Pisum sativum* cultivars that vary in 4C (replicated) DNA content from 13.0 to 161.6 units (Cavallini et al., 1993). There are significant positive correlations between genome size and primary root length 3 days ($r = 0.90$) and 5 days ($r = 0.89$) after germination. There also are significant genome size and stem length correlations 4 days ($r = 0.95$), 7 days ($r = 0.93$), and 14 days ($r = 0.77$) after germination. There are also positive correlations between genome size and differentiated root cell length ($r = 0.86$), epicotyl cell length, and leaf epidermal cell length (Cavallini et al., 1993).

PHYLOGENETIC PATTERNS IN GENOME SIZE

Variation within genera

Congeneric species may differ conspicuously in genome size. In *Ranunculus,* there is a 13-fold difference between a diploid and 16-ploid species (Smith and Bennett, 1975). 4C values range from 7.9 pg in *R. lateriflorus* to 103.2 pg in *R. lingua.* In *Allium,* DNA content per species (from diploids to octaploids) varies across an eightfold range (Ohri, 1998). 4C DNA amounts vary from 35.6 pg in *A. ledebourianum* and 297.1 pg in *A. validum.* In *Nicotiana,* there is a fivefold difference between diploids (4C value of 5.8 pg in *N. trigonophylla*) and tetraploids (29.2 pg in *N. rustica;* Narayan, 1987). In *Crepis,* there is a ninefold difference between the lowest and highest values (Jones and Brown, 1976). The species with the highest base number ($x = 9$) has the smallest genome size.

The effect of polyploidy can be removed by describing the DNA content per genome. It is particularly notable that this value may be highly variable among congeneric species. This is well illustrated in *Allium,* where there is a threefold difference between the smallest and largest values (Ohri, 1998). Polyploids may have less DNA per genome than diploids, as discussed in chapter 8.

There also may be considerable variation in genome size among congeners with the same ploidal level. In *Ranunculus,* there is a fivefold range of values occurs in both diploid and tetraploid species (Smith and Bennett, 1975). The diploid *R. laterifolius* has a 4C value of 7.9 pg, compared with 38.4 pg in diploid *R. ficaria.* The tetraploid *R. sceleratus* has 16.5 pg, compared with 78.5 in the tetraploid race of *R. ficaria.* There is a sixfold difference among diploid *Lathyrus* species (Narayan and McIntyre, 1989). In diploid *Helianthus,* there is a fourfold difference between the lowest (*H. neglectus,* 4C = 12.8 pg) and the highest values (*H. agrestis,* 51.8 pg; Sims and Price, 1985). The South American genus *Bulnesia* has a sixfold difference between diploid species (Poggio and Hunziker, 1986). The genus *Clarkia* has a fourfold difference between diploid species (Narayan, 1988b).

Woody congeners typically have smaller differences in genome size than do herbaceous plants. These differences are thought to be constrained by the maximum nuclear size allowed in small cambial cells of wood fibers (Stebbins, 1950; Khoshoo, 1962). Nevertheless, there are some notable examples of genome size differentiation.

As in herbaceous genera, the range of genome sizes among species is the greatest when species vary in ploidal level. One notable example is in the genus *Terminalia,* where a 3.5-fold difference occurs between the diploid *T. oliveri* (3.6 pg) and the tetraploid *T. bellirica* (Ohri, 1996). Within the diploid species studied, there is 1.97-fold range of variation, and among tetraploid species a 1.76-fold range.

In the genus *Leucena,* 2C levels in diploid species vary from 1.35 pg in *L. esculenta* to 1.81 pg in *L. diversifolia* and from 2.66 pg in tetraploid *L. esculenta* to 3.31 pg in *L. confertifolia* of the same ploidal level (Palomino et al., 1995). The 2C DNA values of 12 *Eucalyptus* species range from 0.77 pg in *E. citriodora* to 1.47 pg in *E. saligna* (Grattapaglia and Bradshaw, 1994). Roughly twofold differences between species of the same ploidal level (diploid) also have been described in arboreal *Cassia* (Ohri et al., 1986) and in *Coffea* (Cros et al., 1995).

The direction of evolution

Except where polyploidy is involved, we have little insight into the direction of genome size evolution during the course of speciation. Progenitor-derivative pairs are useful for such a consideration. One such pair involves the progenitor *Coreopsis nuecensoides* (*n* = 9, 10, 11) and the derived *C. nuecensis* (*n* = 6, 7). The latter has 12% more DNA than the former, even though it has undergone an aneuploid reduction in chromosome number (Price et al., 1984).

Extending the phylogenetic theme, one would like to know the direction(s) of genome size change within a genus. Fortunately, molecular phylogenies are available for a few genera with substantial genome size information.

Cox et al. (1998) have studied a cluster of related slipper-orchid genera (*Paphiopedilum, Phragmipedium,* and *Cypripedium*). *Cypripedium* is characterized by a diploid chromosome number of 20, *Phragmipedium* by numbers from 18 to 30, and *Paphiopedilum* by numbers between 26 and 42. Karyotype analyses suggest that the evolutionary trend has been one of increasing chromosome numbers through centric fission. Thus, all of the species are considered to be diploid. This interpretation is supported by the fact the genome size is unrelated to chromosome number. 4C DNA contents differ 5.7-fold (figure 1.3), from 24.4 pg in *Phragmipedium longifolium* to 138.1 pg in *Paphiope-*

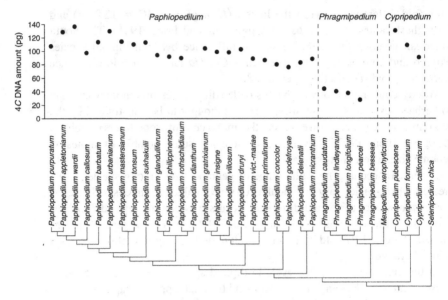

Figure 1.3. Variation in genome size in the Cypripedioideae. Redrawn from Cox et al. (1998), with permission of the Botanical Society of America.

dilum wardii. Species of *Paphiopedilum* have the largest genomes and exhibit the greatest range of variation (71.2–138.1 pg). The phylogenetic study shows a trend toward increasing genome size in *Phragmipedium* and *Paphiopedilum.*

In *Brachyscome,* Watanabe et al. (1999) considered the total length of all chromosomes in relation to a phylogeny based on *matK* sequence data. Total chromosome length varied from 25 μm to 82 μm. Ancestral chromosome length was estimated to be 55.6 μm. A reduction in total chromosome length occurred several times in both subgenera *Brachyscome* and *Metabrachyscome.* However, some lineages of both subgenera displayed increased chromosome length.

Clarkia is another genus with a molecular phylogeny and genome size estimates. Among the diploid species 2C DNA contents vary from 2.16 pg in *C. williamsonii* to 8.87 pg in *C. brewerii* (Narayan, 1988b). Based on the phylogenetic relationships inferred from the nucleotide sequences of the *PgiC* genes (Gottlieb and Ford, 1996), genome size increased with evolutionary advancement.

Phylogenetic considerations

Genome size estimates have proven quite useful in delimiting natural assemblages within herbaceous genera. In *Helianthus,* there is a close association between species that cluster by genome size and by morphology (Schilling and Heiser, 1981) and chloroplast DNA (Rieseberg et al., 1991). Species of *Bulnesia* that apparently are closely related have similar genome sizes (Poggio and

Hunziker, 1986), as do closely related members of *Cicer* (Ohri and Pal, 1991). In trees, DNA contents of *Eucalyptus* (Grattapaglia and Bradshaw, 1994) and *Terminalia* (Ohri, 1996) species conform to their taxonomic classifications.

With the substantial database on angiosperm genome size, it is possible to consider genome size variation within a broad phylogenetic context. One of the first initiatives in this regard was by Bharathan et al. (1994). They found that the presumed early branches within the Ranunculales (the Berberidaceae and Menispermaceae) and the paleoherbs such as Nymphaeales and Piperales have small genomes ranging from 1 to 4 pg/nucleus.

Subsequent phylogenetic analyses by Leitch et al. (1998) provided additional support for a small ancestral genome size in flowering plants (figure 1.4). It is noteworthy that genome sizes exceeding 30 pg/nucleus have arisen in only two major lineages, the monocots and the Santalales. Within these lineages, very large genome sizes are confined to the most derived families. In the monocots, notable family means include the Astromeriaceae (101.0 pg/nucleus), Liliaceae (198 pg/nucleus), and Trilliaceae (244.3 pg/nucleus). None of the Santalales

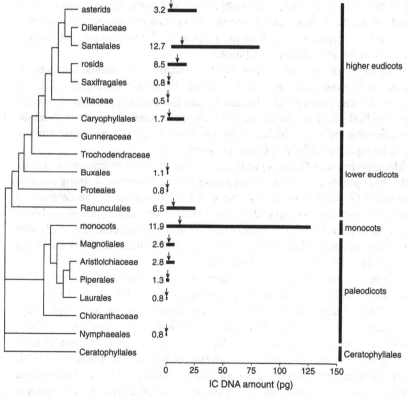

Figure 1.4.   Mean (arrows) and range (lines) of *C*-values for major angiosperm groups in relation to their phylogenetic relationships. Redrawn from Leitch et al. (1998), with permission from the *Annals of Botany*.

even approach these families in genome size. The most notable family is the Loranthaceae, with a mean of 50.9 pg/nucleus.

A phylogenetic analysis of genome size in grasses coupled with a consideration of mechanisms to increase and decrease genome size led Bennetzen and Kellogg (1997) to conclude that angiosperms may have a one-way ticket to genome obesity through amplification of retrotransposons. There is a possibility, however, that genome size could decline through unequal recombination among LTRs of retrotransposons (Rabinowitz, 2000). The longer the LTR, the more likely the transposable element will undergo recombination.

ECOGEOGRAPHICAL PATTERNS IN GENOME SIZE

Avdulov (1931) and Stebbins (1966) observed that species with large chromosomes tend to be concentrated in temperate latitudes. Given that genome size and chromosome volume are closely correlated, we would expect species with large genomes also to be concentrated in temperate areas. This expectation was validated by Levin and Funderberg (1979). The mean 4C DNA value for temperate species (27.06 pg) was more than double that for tropical species (12.13 pg). Significant differences between tropical and temperate species were found within the Poaceae and Liliaceae. Temperate composites and legumes also had larger genome sizes than did their tropical counterparts, but the difference was not statistically significant.

Are differences in genome size due to selection for favorable nucleotypic effects, or do these differences merely reflect the maximum amount of DNA that species can tolerate? If selection targets the nucleotype, then we would expect to find genome size variation within species that correlates with major environmental variables. Such correlations, however, do not prove that genome size is being selected for its phenotypic expression.

Macgillivray and Grime (1995) considered the frost tolerance of British herbaceous plants in relation to their genome sizes. There was a strong positive correlation ($r = 0.55$) between these variables. It remains to be determined whether frost-tolerant herbs in the United States, continental Europe, and Asia also tend to have larger genomes than their less tolerant compatriots. It also remains to be determined whether herbs at higher latitudes tend to have larger genomes than do those at lower latitudes.

In some species, plants in cooler climates tend to have larger genomes. One such species is the grass *Festuca arundinacea*. Ceccarelli et al. (1992) estimated DNA contents for 28 populations from the Italian peninsula. Populations differed by as much as 32%. There was a notable positive correlation between genome size and altitude ($r = 0.60$). Populations in cooler areas tended to have larger genome sizes. The correlation with latitude was less compelling ($r = 0.34$) because altitudinal variation confounded the potential north–south temperature gradient. There was no correlation with precipitation. Differences in genome size were associated with changes in one class of highly repetitive sequences. Larger genome size also was associated with cooler mean January temperatures in Israeli populations of *Hordeum spontaneum* (Turpeinen et al., 1999).

Another example of a positive correlation between genome size and altitude involves the grass *Dasypyrum villosum* (Caceres et al., 1998). Of particular interest are the strong correlations between the genome sizes of individual plants and their altitudes within sites (at the Latium station, $r = 0.74$; at the Basilicata station, $r = 0.82$). The plants with larger genomes had greater total chromosome length that involved all chromosome pairs. Differences in the mean DNA contents of populations had phenotypic correlates, larger values being significantly associated with small leaf length and width, larger seed weight, and longer flowering period. Ostensibly, these differences reflect adaptations achieved through selection on genome size.

Populations with the largest genomes do not always occur at the highest elevations. This is illustrated in French, Spanish, and Italian transects of *Dactylis glomerata* (Reeves et al., 1998), where 1.3-fold variation was observed. The negative correlations between altitude and genome size are striking: France, $r = -0.87$; Italy, $r = -0.82$; Spain, $r = -0.74$. Combining all transects $r = -0.69$ ($p = 0.001$). Parallel variation in the three transects suggests that nucleotypic selection is acting.

Ecogeographical patterns in genome size have been described in several species, two of which are in the composite *Microseris*. The annual *M. bigelovii* is distributed from central California to southern British Columbia and grows on a series of substrates. Price et al. (1981a) reported a 25% range in DNA content. Populations with the lower values occur at the southern extreme of the range, where rainfall is low, and at the northern extreme, where rainfall is high. The larger genome sizes were in the center of the range where rainfall amounts are intermediate.

In *M. douglasii,* there was a 14% variation in genome size among populations in California (Price et al., 1981b). The populations with the most DNA occurred in areas with relatively high rainfall and deep clay soil. Of particular importance is the change in genome size from one year to another (Price et al., 1986). Some populations had increased genome size, whereas others had reduced size. The former grew where the growing season had lengthened; the latter, where it had contracted and become more stressful. To test whether stress can indeed favor lower DNA content, Price et al. (1986) grew six high-DNA plants under severe temperature and water stress in growth chambers. However, the progeny of these plants did not deviate significantly from their parents.

What is the basis for local variation in genome size? Some insights are available from *Hordeum spontaneum.* Kalendar et al. (2000) found a significant positive correlation between elevation and genome size along a 300-meter transect along a single canyon in Israel presenting sharply differing microclimates. Plants in the higher and drier sites had the largest genomes. Their study was noteworthy, because in addition to genome size, they determined the number of copies of the transcriptionally and translationally active BARE-1 (LTR) retrotransposon in each plant. The number varied from 8,300 to 22,100 copies, corresponding to 1.77–4.70% of the nuclear DNA. Of particular interest is the observation that copy number is positively correlated with genome size.

The key issue is the basis for this correlation. Is it a simple by-product of stress-induced retrotransposon activity that varies with the environment, or does BARE-1 activity somehow contribute to local adaptation? Although the answer is unknown, we may be one step closer to understanding whether differences in genome size are adaptive and, if so, the mechanism underlying such.

In addition to the case in *Hordeum,* information is available on particular sequences that vary with genome size among populations of *Vicia faba* from the Mediterranean Basin. Ceccarelli et al. (1995) found interpopulation differences in genome size up to 35%. Tandemly repeated *Fok*1 elements about 60 bp in length were involved in the size differences, the copy number ranging from 5.4×10^6 to 21.5×10^6. These repeats represent up to 9.5% of the *V. faba* genome. The correlation between the copy number and genome size is $r = 0.86$.

The distribution of crop plants and their DNA content shows a pattern similar to that for wild plants. Crops grown in cooler climates (e.g., barley, rye and oats) tend to have larger genomes than those in warmer climates (e.g., sorghum and rice; Bennett, 1976). However, variation patterns within domesticates may or may not follow that trend. Both trends have been reported in *Zea mays* (Rayburn and Augur, 1990).

The key question for maize or any species is whether DNA content will respond to selection or will change in concert with a response to selection on a phenotypic trait. That genome size can indeed be selected was shown in triticale, where a plant lacking the heterochromatin present in the telomeres of the rye genome was established in four generations. In the formulation of this plant, DNA content had declined by 3.5% (Bennett, 1985).

Notably, maize populations selected for cold tolerance had larger genome sizes than did unselected populations (McMurphy and Rayburn, 1991, 1992). The genome size shift was accompanied by the addition of heterochromatic C-bands (the mitotic equivalent of pachytene knobs). DNA content was positively correlated with C-band number (Rayburn et al., 1985). This demonstrates that phenotypic selection can alter genome size. Ostensibly, genome size was altered because it is one of the mechanisms that confers cold tolerance. Genome size also declined in response to selection for early flowering in maize (Rayburn et al., 1994). Apparently, genome size declined because smaller genomes foster faster rates of development.

THE INHERITANCE OF DNA CONTENT

The evolution of genome size is predicated on the heritability of DNA content. If differences among plants were due to environmental factors alone, then a response to selection would be impossible.

Only a few studies have dealt with the inheritance of DNA content. As discussed below, DNA content does not follow the pattern one normally finds for a quantitative trait. One of the first studies on the inheritance of DNA content was conducted on *Lolium* species and their hybrids. Hutchinson et al. (1979) made crosses between outcrossing species with relatively large 2C DNA contents (*L. remotum,* 6.04 pg; *L. temulentum,* 6.23 pg) and three inbreeders with relatively

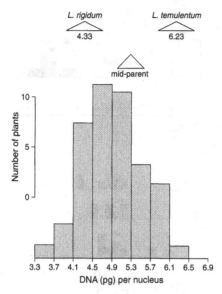

Figure 1.5. Distribution of DNA values per 2C nucleus among the F_2 progeny of the cross *Lolium temulentum* × *L. rigidum*. Redrawn from Hutchison et al. (1979), with permission of Blackwell Science Ltd.

small contents (*L. perenne*, 4.16 pg; *L. multiflorum*, 4.31 pg; *L. rigidum*, 4.33 pg). The F_2 progeny of the cross between *L. temulentum* and *L. rigidum* displayed segregation of genome size, with a mean significantly below the parental mid-point (figure 1.5). The F_2 generation from the *L. remotum* × *L. rigidum* cross also deviated significantly in the direction of the species with the smaller genome. Conversely, the F_2 generation from the *L. temulentum* × *L. multiflorum* cross did not differ significantly from the mid-parent value. However, backcrosses to *L. multiflorum* had less than the expected amount of DNA.

This study is of particular interest because Hutchinson et al. (1979) considered several vegetative and reproductive expressions of adult F_2 generations in relation to their DNA content. Surprisingly, there were no correlations in two F_2 populations and a near absence of correlations in the third.

Price et al. (1983) analyzed the products of crosses between *Microseris douglasii* and *M. bigelovii*, which differ in DNA content by about 10%. The 12 F_1 progeny were distributed nearly all across the spectrum between the parental values (figure 1.6). The F_2 families derived by selfing different F_1 progeny had means similar to their respective parents. In a later study involving different members of *M. douglasii* and *M. bigelovii*, Price et al. (1985) found that the F_1 hybrids tended to have genome sizes similar to those of the paternal parent, as opposed to the expected intermediate value. However, the F_2 families tended to revert back toward the values of the maternal species, rather than retaining the value of the F_1 progeny from which they were derived (figure 1.7). More-

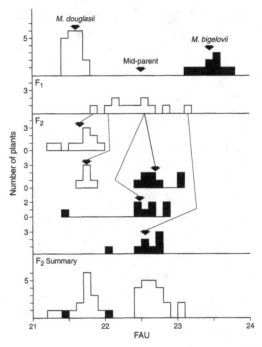

Figure 1.6. DNA contents (Feulgen Absorbancy Units, FAU) per 2*C* nucleus in *Microseris douglasii, M. biglovii,* F$_1$ hybrids, and five F$_2$ families derived from different F$_1$ plants. Redrawn from Price et. al. (1983), with permission of the Botanical Society of America.

over, the segregation of DNA values expected in the F$_2$ families was largely absent. The few F$_3$ families studied had means similar to their F$_2$ parents.

The patterns of genome size inheritance in *Microseris* suggest that genome size in hybrids is unstable, being subject to deletion and amplification. Price et al. (1985) proposed that the changes occurred sometime between fertilization and when the leaves used for analysis were formed. Once altered, the change may be stabilized and persist in the subsequent generations.

Evidence for instability in genome size also comes from maize. Rayburn et al. (1993) determined the nuclear DNA content of 9 inbred lines and 14 hybrid combinations. In some combinations, siblings had a broad spectrum of genome sizes, even surpassing the sizes of their parents. In other combinations, siblings were clustered, sometimes around a value that was substantially larger than the mid-parent value. Indeed, in five parental combinations, the mean nuclear DNA content of the F$_1$ hybrids significantly exceeded the parental means.

Another example of F$_1$ hybrids deviating in DNA content from the mid-parent value was reported in flax. Durrant (1981) crossed plants of the variety Stormont Cirris with high and low DNA contents (16% difference) with the

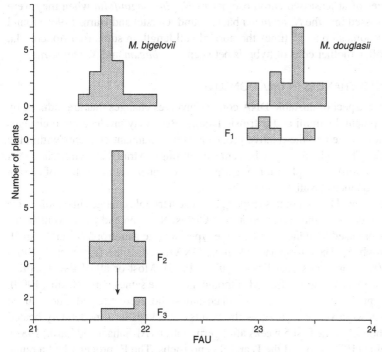

Figure 1.7. DNA contents (FAU) per 2*C* nucleus in *Microseris douglasii, M. biglovii,* F$_1$ hybrids, and a F$_2$ family derived from the F$_1$ plant indicated. Redrawn from Price et. al. (1985) with permission of Urban and Fischer Verlag.

variety Liral Monarch, which had an intermediate DNA content. The F$_1$ progeny had DNA contents similar to that of Liral Monarch, as opposed to the expected intermediate levels, signifying that a quantitative change in DNA content occurred in these hybrids.

Differential DNA replication may be the basis for the deviation of hybrids from the parental mid-point. Consider the hybrid between *Helianthus annuus* and *H. tuberosus* (Natali et al., 1998). The former (2*n* = 34) had a mean 4C DNA content of 21.5 pg versus 50.0 in the latter (2*n* = 102). First generation hybrids had a mean DNA of 42.0 pg, significantly above the mid-parent value. The hybrids had 68 chromosomes as expected. Thermal denaturation of genomic DNA showed that in hybrids some DNA families differed in redundancy relative to that expected from parental expressions. Differential DNA replication also was reported for hybrids between *Nicotiana rustica* and *N. tabacum* (Neelam and Narayah, 1994) and hybrids between *Triticum aestivum* and *Secale cereale* (Lapitan et al., 1988).

Cytological information pointed to unstable DNA content in *Nicotiana* many years ago. Moav et al. (1968) reported marked heritable increases (~50%) in the sizes of at least two chromosomes of *N. plumbaginifolia* when they were backcrossed into the *N. tabacum* background. Gerstel and Burns (1966) found chromosomes up to 15 times the normal cell length in some root tip, corolla, and pollen mother cells of hybrids between *N. tabacum* and *N. otophora*.

INDUCED CHANGES IN DNA CONTENT

Another aspect of unstable DNA content involves changes that are induced by developmental stimuli and extrinsic factors. Both may produce rapid changes in genome size and the relative proportions of redundant elements within the genome. These changes may be achieved through extra DNA synthesis (DNA amplification), nonreplication of certain DNA sequences, or the loss of certain DNA sequences (Walbot and Cullis, 1985).

Flax is well known in this respect, because heritable changes in genome size can be induced in the variety Stormont Cirrus. Nitrogen and phosphorus fertilizers were used to induce two extreme types: a large genotroph (L) and a small genotroph (S). The L form had 16% more DNA than did the S (Evans et al., 1966) and 50% more ribosomal DNA (Cullis, 1976). Most or all of the repetitive sequence families were affected, although not to the same degree (Cullis, 1990). Not surprisingly, then, the sizes of all chromosomes were affected. The induced changes were stable as long as the plants were kept in a moderately heated greenhouse for the first 5 weeks after germination (Al-Sahael and Larik, 1985).

Cullis (1979) crossed the L and S genotrophs. The F_1 progeny had a range exceeding the parental values in rDNA numbers, which differed by about 50%. The extreme F_1 progeny were selfed, and their progeny again yielded a range similar to that present in the F_1 offspring. Selfing a series of F_2 progeny produced F_3 families whose mean rRNA gene numbers were not correlated with the F_2 values. Thus, ribosomal gene number is not only sensitive to the environment but also unstable in its transmission to the next generation.

Day length also may affect genome size. Dhillon (1988) reported that the DNA content of *Populus deltoides* leaves was 27% higher in plants grown under long days (12 hrs) than plants grown under short days (9 hrs).

Another extrinsic factor that may affect genome size is temperature. For example, seeds of *Festuca arundinacea* germinated at 10°C, 20°C, and 30°C had increasing genome sizes in each of seven populations (Ceccarelli et al., 1992). Conversely, genome sizes did not undergo changes when plants were exposed to temperature differences at later developmental stages. It remains to be determined whether the changes induced by temperature are heritable.

Temperature also has been shown to affect the copy number of rDNA in *Brassica nigra*. Waters and Schaal (1996) subjected plants to a heat shock of 40°C with high humidity for 2 hr. This treatment reduced the copy number by 37% relative to control plants. Moreover, the reduced copy number was heritable.

Consider next how a developmental variable may affect genome size. Natali et al. (1993) analyzed progeny in a *Helianthus annuus* line, which was the product of 10 generations of self-fertilization. Progeny from single individuals were expected to be uniform in DNA content, given that the plants were highly inbred. This, however, was not the case. Members of the same family varied in DNA content. Much of the variation stemmed from the position of the achenes in flowering heads from which plants arose. The nuclear DNA content of seedlings from achenes drawn from the periphery of heads was 14.7% higher than that of seedlings from achenes drawn from the center of heads.

DNA sequences were broken down by thermal denaturation and partitioned from their reassociation kinetics into highly repeated, fast medium repeated, and intermediate medium repeated fractions. The seedlings from the periphery had 1.40 times as many highly repeated copies, 1.34 times as many fast medium copies, and 2.14 times as many intermediate medium copies as the central seedlings.

Plants from peripheral and central achenes differed in several respects, which are consistent with the expected nucleotypic effects. Plants from peripheral achenes had larger cells, longer mitotic cycle times, and a greater number of days to flowering than those from central achenes.

Another example of differences in DNA content associated with divergent seed types is in *Dasypyrum villosum*. Individual plants have brown and yellow caryopses. Frediani et al. (1994) found that 2-day-old seedlings from yellow caryopses had 12% more DNA than those of the brown caryopses. Moreover, DNA content increases in each seed during germination, to a greater extent in the seedlings of yellow caryopses than in those from the brown caryopses. The researchers estimated the copy number of sequences related to a 396-bp repeat in resting embryos and seedlings of both caryopsis types. These repeats were distributed among six of the seven chromosome pairs. The copy number underwent a temporal change similar to that of total DNA and was much greater in the yellow seeds and seedlings. Thus, there exists a "fluid" domain in plant DNA. The level of sequence redundancy may be one mechanism by which plants regulate their development.

OVERVIEW

Given what we know about correlations between DNA content and the phenotype, ecogeographical patterns in DNA content within and among species, the inheritance of DNA content, and environmental alteration of DNA content, we may revisit the two opposing viewpoints on the significance of genome size brought about by differences in noncoding DNA. One viewpoint is that noncoding DNA essentially is "junk" DNA. Its only function is self-preservation and growth, somewhat like a parasite. Its alteration has a negligible effect on the organism. A second view is that repetitive DNA has a functional and thereby adaptive role. Nuclear DNA thus may influence the phenotype through the expression of its genic content and also by the physical effects of its volume and mass.

For the adaptive view to be valid, there must be evidence that (1) change in genome size causes phenotypic change, (2) differences in genome size are heritable, and (3) phenotypic selection on traits related to fitness can alter DNA content. As discussed in this chapter, there is ample evidence for a link between genome size and the phenotype and for the heritability of genome size, and some evidence that phenotypic selection can be accompanied by genome size change. Moreover, we know that the placement of structural genes with respect to heterochromatic chromosomal regions (i.e., those containing repetitive sequences) affect gene expression (Li and Grauer, 1991; Zuckerkandl and Hennig, 1995). Thus, a strong, if not conclusive, argument can be made in favor of an adaptive role for genome size.

. It is clear that the plant genome is not quantitatively static during growth. As endopolyploidy affects morphogenesis and the functional activity of the organism, so may substantial amplification, silencing, or loss of repetitive sequences or their transposition. Walbot and Cullis (1985, 368) advocate an adaptive role for induced genomic change: "Genomic flexibility is an aspect of the extraordinary adaptability of plants to a changing environment." Genome flexibility generates genetic variants that may be better able to cope with environmental stress.

In favor of the position of Walbot and Cullis (1985) is the fact that the changes induced by stress are specific to certain parts of the genome and are repeatable. However, the predictable nature of the changes does not in and of itself prove that they are adaptive, nor does the fact that the changes may affect the phenotype. The critical experiments demonstrating that induced genomic change leads to enhanced fitness have yet to be performed.

2

Chromosomal Rearrangements within Species

Many plant species are heterogeneous for translocations, inversions, fusions, and fissions. These rearrangements are likely to alter the size, banding pattern, and symmetry of chromosomes (i.e., the ratio of the short to long chromosome arms). Their specific nature may not be evident even if meiotic pairing can be studied in chromosomal heterozygotes. Chromosomal heterozygotes are likely to have reduced fertility depending on the magnitude and number of rearrangements and, in the case of inversions, the pattern of crossing over. The aforementioned rearrangements need not affect chromosome number, although they may cause this number to increase or decrease.

The purpose of this chapter is to explore the spatial patterns of chromosomal variation and the conditions that allow the fixation of novel variants. Fixation establishes a barrier or retardant to gene exchange between populations and thus is important in allowing populations to pursue independent avenues of evolution in the face of pollen and seed exchange.

Before proceeding, it is useful to review the meiotic signatures of translocation and inversions in heterozygotes, because these are the most common rearrangements. Chromosomes involved in a reciprocal translocation between two nonhomologous chromosomes typically form quadrivalents (rings or chains of four chromosomes). If a chromosome is involved in more than one translocation, the multivalents may involve more than four chromosomes. If a chromosome has undergone an inversion in which the centromere is not involved (paracentric inversion), a heterozygote for this condition will have a dicentric bridge and acentric fragment at anaphase I, if there is chiasma formation in the inverted region during prophase I. Inversions that involve the centromere (pericentric inversions) typically do not have meiotic irregularities, but like translocations and paracentric inversions, they will cause reduced fertility in a heterozygous condition.

The level and geographical organization of chromosomal heterogeneity vary widely among species. Rearrangements may be anywhere from rare to common, and they may be distributed randomly in space or have some spatial organization.

TRANSLOCATION HETEROZYGOSITY

Heterozygosity levels

The genus *Clarkia* is notorious for translocation heterozygosity and, indeed, has been a model genus for the study of chromosome evolution. The cytology of several species of *Clarkia* has been studied quite intensively by Harlan Lewis and his associates.

In the diploid *C. williamsonii* ($2n = 18$) many, perhaps most, populations have a significant level of chromosomal heterozygosity. Wedberg, Lewis, and Venkatesh (1968) analyzed prophase I configurations in plants from 30 populations distributed along an elevational gradient from 300 to 1700 meters in the Sierra Nevada mountains of California. Some plants exhibited single rings of four chromosomes, others two rings of four chromosomes each, and others single rings of six chromosomes. Of the 963 plants studied, 298 were heterozygous for one translocation, and 31 were heterozygous for two translocations. In only one instance was the same arrangement found in two populations. The translocation rings had normal alternate disjunction and thus were not involved with a reduction in fertility.

The frequencies of heterozygosity within populations varied from 0.02 to 1.0. The distribution of population frequencies is depicted in figure 2.1. Populations above 1000 meters have values between 0.02 and 0.25, whereas nearly all of those at lower elevations have values exceeding 0.25. There is no linear relationship between elevation and heterozygosity.

Translocation heterozygosity also is prominent is *C. amoena* ($2n = 14$). Snow (1963) examined 197 plants from 37 populations and found that 62% of the plants were heterozygous for one or more translocations. All but two populations had translocation heterozygotes. The standard end arrangement prevailed in most populations. Some populations were unique in their derived end arrangements, whereas others shared derived arrangements. There is no evidence that the existing polymorphism has given rise to distinctive chromosome races.

C. unguiculata ($2n = 14$) is another member of the genus that has frequent and widespread translocation heterozygosity. Heterozygotes comprised over 35% of the plants studied from 36 populations (Mooring, 1958). In some populations, heterozygotes are nearly as frequent as homozygotes. In addition to the normal arrangement, four novel end arrangements are present in homozygotes in sporadic localized populations near the ecological limits of the species. This is similar to the overall pattern in its close ally, *C. exilis,* in which the most

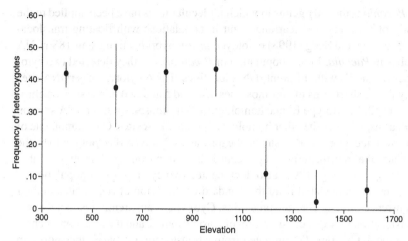

Figure 2.1.   The frequency of translocation heterozygotes in relation to elevation in *Clarkia williamsonii*. Circles denote population means, and lines denote 95% confidence limits. Redrawn from Wedberg et al. (1968), with permission of the Society for the Study of Evolution.

common arrangement is found throughout the range and novel homozygotes occur in ecologically marginal populations (Vasek, 1960).

Translocation heterozygosity also is common in *C. elegans* ($2n = 16$), appearing in 35% of the plants sampled from 16 populations (Lewis, 1951). Heterozygotes were present in 8 of the 16 populations. Lewis identified five novel end arrangements resulting from translocations and two others resulting from inversions.

In *C. dudleyana* ($2n = 18$), most populations exhibit little translocation heterozygosity, although in one population near the center of the range, 69% of the plants sampled were heterozygous (Snow, 1960). Hybrids between populations have meiotic configurations connoting that they differ in one to several translocations. However, owing to alternate disjunction of the chromosomes in the rings, hybrids with rings of four or six chromosomes are highly fertile. The species thus has moved beyond the stage of translocation heterozygosity. Several rearrangements have been fixed in different populations but have yet to spread very far.

Paeonia is another genus in which some species have extensive translocation heterozygosity. The chromosome configurations in *P. californica* ($2n = 10$) include all combinations of pairs and rings possible with 10 chromosomes (Walters, 1942). There is a tendency for plants growing near the center of the distribution to be more homozygous than those near the limits of the range. Individual populations typically show a range of configurations; the most common one is shown in figure 2.2. Heterozygotes are not maintained by a balanced lethal system. Rings containing from four to eight chromosomes also have been described in *P. brownii* (Grant, 1975).

Paeonia is the only genus in which molecular tools have been applied to the study of chromosomal rearrangement in populations with floating translocations. Zhang and Sang (1998) employed *in situ* hybridization using 18S rDNA probes in *Paeonia.* In one population of *P. californica,* they detected two cytotypes (A and B) with different rDNA locations. The A cytotype has rDNA sites only on the short arms of chromosomes 3, 4, and 5 and two minor sites on chromosome 2. In cytotype B, one homologue of chromosome 4 has rDNA sites on both arms, whereas the other homologue has no rDNA sites. One homologue of chromosome 3 has a rDNA site on the long arm. Cytotype B is thought to be the product of a translocation and pericentric inversion from the A arrangement.

Zhang and Sang (1998) also documented two cytotypes in a population of *P. brownii* that deviated from the standard organization of rDNA sites in their occurrence on long chromosome arms. Cytotype C apparently resulted from a translocation between the short arm of chromosome 2 and the long arm of chromosome 4. Cytotype D originated from a translocation between the short arm of chromosome 3 and the long arm of chromosome 4.

The extensive chromosomal heterozygosity in *Clarkia* and *Paeonia* is quite unusual. A more representative situation is that in *Allium schoenoprasum* (Liliaceae). Stevens and Bougourd (1991) analyzed meiotic configurations in 1,017 plants from 18 populations in Europe. Twenty-three plants were heterozygous for translocations, and 12 were heterozygous for paracentric inversions. Two plants had two translocations, the rest one. Nineteen different translocations and at least two different inversions were identified.

In the diploid grass *Alopecurus aequalis,* two of seven populations studied had translocation heterozygotes (Sieber and Murray, 1981). In one population

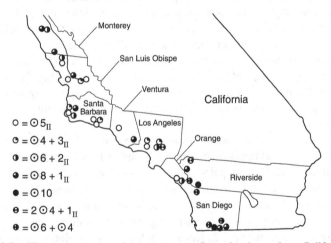

Figure 2.2.   The most common chromosome configuration in southern California populations of *Paeonia californica.* Redrawn from Walters (1942), with permission of the Botanical Society of America.

Table 2.1. The incidence of chromosomal heterozygosity in diploid *Aloe* and its allies.

Rearrangement	Genus		
	Aloe	*Haworthia*	*Gasteria*
Paracentric inversion	6.5%	0	4.2%
Pericentric inversion	0.2%	0.2%	0
Translocation	2.0%	3.9%	1.5%

Source: Adapted from Brandham (1983).

(Ruislip), heterozygosities were 22%, 20%, and 10% in 1976, 1977, and 1978, respectively. In the other population (Mislen), 30% of the plants were heterozygotes. Pollen fertility in heterozygotes was reduced, varying from 55% to 95%. Approximately half of the progeny obtained from selfing a heterozygote were heterozygous, as would be expected if chromosome disjunction from a ring of four chromosomes was random, that is, alternate disjunction half of the time and adjacent disjunction the other half.

Structural heterozygosity even occurs in taxa embedded in groups that display very little chromosomal diversity. Consider the Aloaceae, whose three major genera (*Aloe, Gasteria,* and *Haworthia*) and the few minor ones all have the same basic chromosome number ($x = 7$) and have one long submetacentric, three long acrocentric, and three short acrocentric chromosomes (Brandham, 1983). The incidence of inversion and translocation heterozygosity in large samples of these genera is surprisingly high (table 2.1), given the conservative nature of the karyotype. Of particular interest is the absence of novel homozygotes despite the appearance of heterozygotes. Brandham concludes that they must be a selective liability and eliminated very early in development. It may be that the karyotype per se is under selection (Brandham and Doherty, 1998).

The genesis and retention of heterozygosity

What factors account for the unusual levels and varieties of translocation heterozygosity in *Clarkia* and in other genera? Relatively high rates of spontaneous chromosomal breakage and heterozygote advantage appear to be the prime causal agents. Lewis and Raven (1958) propose that there may be a heritable attribute underlying chromosome breakage. Unfortunately, there have been no direct measures of spontaneous breakage rates. Fragments indicating breakage seem to be very rare. Frequent chromosome breakage probably accounts for high levels of translocations in some populations of *Paeonia californica* (Walters, 1956) and diverse karyotypes in small populations of *Rumex acetosa* (Parker and Wilby, 1989).

The unusual levels of heterozygosity in *Clarkia,* all of which are outcrossers, may lie in the preservation of genic heterozygosity during sporadic population contractions and attendant inbreeding (Wedberg et al., 1968). When heterozygous, reciprocal translocations prevent recombination between the centromeres and the points of breakage and between the two break points in those chromosomes involved in more than one translocation (Darlington, 1936). Chromosomal heterozygotes then would be expected to be more vigorous and better able to contend with environmental stress than chromosomal homozygotes.

Evidence in support of heterozygote superiority is forthcoming from interracial hybrids of *Clarkia speciosa*. These hybrids produce rings ranging in size from 4 to 10 chromosomes. Bloom (1976) selfed one heterozygote with a ring of 4, two with rings of 8, and one with a ring of 10 chromosomes. He recovered structural homozygotes and heterozygotes in the progeny and compared their dry weights after growth in a common controlled environment. The heterozygotes with the rings of four and rings of eight were roughly 50% larger than their homozygote counterparts.

Additional evidence of heterozygote advantage comes from *Campanula* and *Secale*. Translocation heterozygotes are common in natural populations of the normally outcrossing *Campanula persiciflora* (Campanulaceae). When these plants are self-fertilized in the greenhouse, none of their progeny are structural homozygotes (Darlington and La Cour, 1950). In rye, which also is an outcrosser, selfing translocation heterozygotes leads to progeny arrays in which the heterozygote may exceed the expected incidence of 50%. The deviation depends on the line and the translocation but is not as pronounced as in *Campanula* (Rees, 1961).

The rye study was extended to include competition among progeny of selfed translocation heterozygotes over a range of densities from 1 to 150 plants per pot (Bailey et al., 1976). Mortality before flowering increased from 40% with one plant per pot to over 90% at the highest density. As the mortality rate increased, so did the percentage of survivors that were translocation heterozygotes (figure 2.3).

INVERSION HETEROZYGOSITY

The documentation of inversion polymorphism is best made from pachytene analyses where the tell-tale loop configuration may be observed. Most often such analyses are not made, and reliance typically is placed on the formation of bridges and fragments in anaphase I that are formed from crossing over in paracentric inversions. The problem with this approach is that bridges and fragments may be the products of chromosome breakage and reunion as well as inversion heterozygosity. In some species (such as maize), there is a strong correlation between the frequency of bridge-and-fragment formation and the frequency of pachytene loops (Maguire, 1966). In other species with relatively large chromosomes (e.g., *Podophyllum peltatum*), anaphase bridges form in the absence of pachytene loops (Newman, 1966). If we assume that all bridge-and-fragment configurations are manifestations of inversion heterozygosity, we are likely to overestimate the level of inversion heterozygosity in plant populations.

With caution in mind, it is useful to consider some data on bridge-and-fragment configurations. These configurations, perhaps indicative of inversion

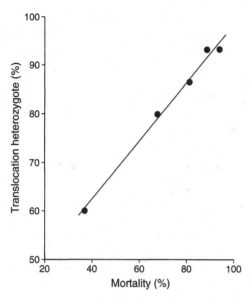

Figure 2.3. The relationship between the percentage translocation heterozygotes and percentage mortality in rye cultures undergoing density stress. The percentage of heterozygotes is assumed to be 50% in each culture at the time of planting. Redrawn from Bailey et al. (1976), with the permission of Blackwell Science Ltd.

heterozygosity, are most conspicuous in *Paris quadrifolia* (Geitler, 1937, 1938) and *Paeonia californica* (Stebbins and Ellerton, 1939; Walters, 1942, 1952). Indeed, they were observed in almost every plant examined.

Bridges and fragments also occur in *Atriplex longipes* (Gustafsson, 1972). There is a negative correlation between bridge formation and pollen fertility (figure 2.4). This species is interesting because in populations with fewer than 100 plants, about a quarter of the plants have pollen fertilities less than 80%. The normal range is between 80% and 100%. This pattern may mean that small populations have relatively high levels of inversion heterozygosity. Moreover, crosses between some populations in which fertility is normal yield progeny with much reduced fertility, which may indicate that populations are fixed for different inversions.

A somewhat similar situation exists in *Anemone multifida* and *A. tetonensis* (Madahar and Heimburger, 1969). Populations display a very low incidence of bridge-and-fragment formation. Between zero and 23% of the pollen mother cells (PMCs) from progeny of conspecific interpopulation crosses have such configurations, and from about 45% to 55% of the PMCs in the progeny of interspecific crosses have bridges and fragments. There was a negative correlation between the incidence of meiotic abnormalities and pollen fertility.

Pericentric inversion heterozygosity also was reported in *Calycadenia ciliosa* (Carr and Carr, 1983). This occurrence probably is the result of crossing between

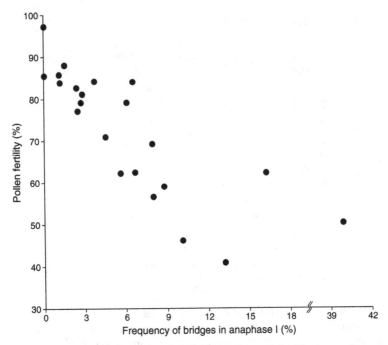

Figure 2.4. The relationship between pollen fertility and the frequency of anaphase I bridges in *Atriplex longipes*. Redrawn from Gustafsson (1972), with permission of *Hereditas.*

two chromosomal races, rather than arising de novo within populations. Crosses between chromosome races yield hybrids with looplike configurations involving two chromosomes.

Pericentric inversion heterozygosity does not leave an anaphase signature. However, pericentric inversions may alter the symmetry of chromosomes and lead to heteromorphic chromosome pairs at diakinesis. Walters (1952) observed such pairs in *Paeonia californica.* He identified two distinct types. One was found in 22 plants in over 500 examined. These plants were scattered over 10 widely separated localities. The other heteromorphic type occurred in only one plant. Heteromorphic pairs occurred in individuals with only bivalents as well as those with ring multiples.

CHROMOSOMAL HETEROZYGOSITY IN MARGINAL POPULATIONS

Geographically marginal populations are more chromosomally variable and heterozygous than interior populations in many species. In *Isotoma petraea*, translocation homozygosity gives way to partial and eventually complete heterozygosity in the southwestern corner of the species range (James, 1965). Similarly, in *Paeonia californica*, plants with bivalents or only small rings occur near the center of the range, whereas complete translocation heterozygotes are located near ecological and geographical margins (Walters, 1942). Translocation heterozygotes in *Clarkia williamsonii* are common in marginal populations but are rare in the center part of the range (Wedberg et al., 1968). Higher levels of chromosomal heterozygosity also has been described in peripheral populations of *Haworthia reinwardtii* (Brandham, 1974) and is especially pronounced in small, isolated, peripheral populations of *Atriplex longipes* and *A. triangularis* (Gustafsson, 1972, 1974), *Elymus rechingeri* (Heneen and Runemark, 1962), *E. striatulus* (Hennen, 1972), *Erysimum* sect. *Cheiranthus* (Snogerup, 1967), *Allium* spp. (Bothmer, 1970), and in *Leopoldia comosa* and *L. weissii* (Bentzer, 1972a,b).

In the dioecious *Rumex acetosa,* a small population on the small island of Skomer (off the coast of Wales) has 14 unique chromosome arrangements (Parker and Wilby, 1989). Eleven of them reside in 67 plants. Chromosomal heterozygosity is much greater on the island than on the adjacent mainland. The unusual level of chromosomal variation is associated with a severe bottleneck in population size.

Elevated levels of chromosome breakage related to inbreeding probably is most important in causing higher levels of polymorphism and heterozygosity in marginal populations. The aforementioned *Rumex* population recently had undergone a severe contraction. Many populations referred to above with relatively high levels of heterozygosity are relatively small and isolated and thus prone to continuous inbreeding that may be amplified during periodic contractions.

What is the evidence that inbreeding per se can promote chromosome breakage? The frequency of spontaneous chromosomal rearrangement in normally outbred taxa such as rye (Jones, 1969; Hrishi, 1969) and *Alopecurus* (Johnsson, 1944) increases with enforced inbreeding. In maize, inbreeding increases the incidence of radiation-induced chromosomal breakage (Stoilov et al., 1966). The situation is similar to that in animals, where inbreeding in the grasshopper *Pyrgomorpha kraussii* is accompanied by a burst of structural rearrangements in the germline (Lewis and John, 1959).

If the origin of novel chromosomal variants were related to inbreeding, then small populations would be expected to accumulate more differences than large populations. Data bearing on this expectation have been obtained by Gustafsson (1974) for *Atriplex longipes* ssp. *praecox.* The fertility of progeny from crosses between small populations tends to be less than the fertility of progeny between large populations. The distributions of fertilities for the two sets of crosses is shown in figure 2.5. Meiotic analyses indicate that inversion heterozygosity is the primary fertility depressant.

Based on the discussion above and the literature in general, inversion heterozygosity apparently is much less frequent than is translocation heterozygosity. This difference may be real or an artifact of the relative ease in detecting translocations. Perhaps translocations simply occur with a higher frequency than do inversions. Whittingham and Stebbins (1969) addressed the latter issue in an intriguing way, using as their subject *Plantago insularis,* whose pachytene

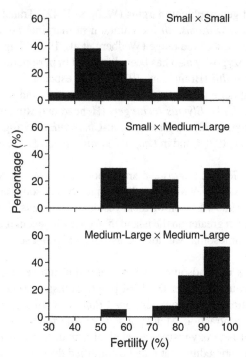

Figure 2.5. The distribution of mean pollen fertility values in crosses between *Atriplex longipes* populations in relation to population size. Adapted from Gustafsson (1974).

chromosomes lend themselves to careful scrutiny. They exposed presoaked seeds to 20,000 roentgens of gamma irradiation, which yielded 64 semisterile plants. Twenty-four of these plants were heterozygous for one or more aberrations; 67% were translocations and 33% were inversions. Gottschalk (1951) conducted a similar study on *Solanum lycopersicon* and found that 89% of all aberrations were translocations.

In both studies, most chromosome breaks occurred in heterochromatic regions. In *Plantago,* most breaks near the centromere resulted in translocations, whereas the majority of breaks away from the centromere yielded inversions. Translocations may prevail over inversions because exchange takes place only between damaged chromosomes that are in close proximity. The probability of contact between heterochromatic regions of different chromosomes is greater than that between heterochromatic regions of the same or homologous chromosomes.

THE FIXATION OF NOVEL CHROMOSOME ARRANGEMENTS

Lande (1979) estimated that the spontaneous rates of occurrence of different types of chromosomal rearrangements in animals are around 10^{-4} to 10^{-3} per gamete per generation, with reciprocal translocations tending to be more common than inversions when the number of chromosomes is large. The rate of chromosomal rearrangement in plants may be higher, at least in some groups. Brandham (1983) measured the frequency of structural change in hundreds of gametes from normal plants in *Haworthia* as evident in viable progeny. He found that the rate was close to zero for gametes of diploid plants. However, about 0.45% of the gametes of tetraploids carried translocations, and a similar percentage carried inversions.

Whatever the rate of chromosomal rearrangement, each novel translocation and inversion initially is rare and occurs in a heterozygous condition. Being in a heterozygous condition imposes a liability on the rearrangement, because heterozygous plants have reduced fertility. Heterozygotes for all but the smallest translocations typically have about 50% fertility. The higher fertility in *Clarkia* heterozygotes is atypical and achieved because chromosome disjunction at anaphase I tends to be alternate. The fertility of inversion heterozygotes depends on the frequency of crossing over in the inverted segment. Most novel arrangements have a high probability of being lost from a population, even though they may not be disadvantageous in a homozygous state.

There has been considerable interest in the mechanism that would allow the fixation of novel chromosomal arrangements. How is heterozygote disadvantage (or underdominance) overcome? The process of establishment of an underdominant rearrangement has been studied from a theoretical perspective by several authors in a variety of models. To account for the fixation of major rearrangements that are underdominant, one must postulate either a deterministic advantage sufficient to overcome heterozygote disadvantage (e.g., meiotic drive or linkage with favorable gene combination) or stochastic processes (Lande, 1985). The latter seems to be the most likely factor.

Wright (1940, 1941, 1969) recognized that populations with small effective sizes facilitate the fixation of underdominant rearrangements through genetic drift. According to Wright, the effective size of a population through time, Ne, is the number of breeding adults in an ideal population of constant size that would undergo the same rate of genetic drift as the actual population when gametes are sampled at random. When taking into account short-term fluctuations in population size, Ne is approximately equal to the harmonic mean of the effective sizes in the separate generations. Because small values predominate in harmonic means, the effective population size over time will be closer to the minimum value than to the average. For example, if in five successive generations the Ne values were 500, 10, 500, 500, and 500, the harmonic mean would be 46.3, as opposed to the mean of the values, which is 402.

The effective size of a population is likely to be much less than actual size (N), if there are wide fluctuations in population size or if populations have recently been founded, especially by a few individuals (Wright, 1969). If the individuals were related, the effective size is depressed even further relative to the actual size (Hedrick and Levin, 1984). This is most likely to occur in plants where multiseeded fruits are the unit of dispersal.

The effective size of a population also is depressed by inbreeding. Thus, predominantly selfing species are expected to have smaller effective sizes than are outcrossers, all else being equal (Hedrick, 1981). Empirical support for this expectation is forthcoming from Ne estimates based on allozyme frequencies (Schoen and Brown, 1991).

Another factor acting to reduce the effective size is high variance in reproductive success (Wright, 1938a; Nunney, 1991). Plants are especially notable in this respect. The distributions of plant fecundities are rather L-shaped in many populations, with a very small proportion of the population making a substantial and grossly disproportionate contribution per capita to the seed output of the population (Levin, 1978). Based on observed fecundity distributions alone, Heywood (1986) showed that effective sizes in annual plants may be well below 40% of the actual size.

Still another factor that affects the effective size of populations is a persistent seed bank. Some seeds of many species, especially annuals, may remain in the soil for many years before they germinate. A well-developed long-lived seed bank increases the effective size of populations and reduces the effect of stochastic processes on genic and chromosomal frequencies (Templeton and Levin, 1979).

Based on the dynamics of plant population size and reproduction, the effective sizes of populations, especially annuals with limited seed banks, are likely to be very much less than the number of reproductive adults. In some species, Ne/N may be 0.1 or less.

A prime effective size estimation based on more than one parameter involves a population of *Papaver dubium* that contained 2,316 plants. Fifty percent of all seed were produced by 2% of the most fecund plants, with the most fecund plant producing 4.6% of the seeds (Mackay, 1980; Crawford, 1984). This highly nonrandom variation in fecundity yields a ratio of $Ne/N = 0.07$. About 75% of the seeds produced are products of self-fertilization. Taking this into account, the ratio of Ne/N is reduced to 0.024 (Gale, 1990).

The effective population size is important because the probability of fixing a novel underdominant rearrangement in a local population is an inverse function of the effective size. There are two issues to consider regarding such fixation. One is the probability of fixation in a single local population starting with a single mutant. The other issue is the rate of fixation per generation.

The probability of fixing an underdominant rearrangement has been studied quite completely (Walsh, 1982; Lande, 1985). The probability of fixation is a negative function of the effective size and heterozygote fitness. The probability of a novel rearrangement being fixed with heterozygote relative fitness of 0.5 (the typical case for translocations) and effective population size of 40 is 5×10^{-6}. Holding population size constant, a relative fitness of 0.90 yields a fixation probability of 5.6×10^{-4}, and a relative fitness of 0.98 yields a fixation probability of 0.019. If the heterozygote and both homozygotes have equal fitnesses, then the probability of fixation will be the same as that of a neutral mutation, $1/2N$.

These estimates assume that population size remains small and that the population receives no immigrants. Rapid population growth following a bottleneck or founding episode (Hedrick, 1981) or immigration (Lande, 1985) would decrease the fixation probability. Conversely, if the novel arrangement were favored in a homozygous state, the fixation probability would be increased.

As discussed above, an increase in heterozygote fitness from 0.5 to 0.9 increased the probability of fixation of a new chromosomal arrangement 100-fold. What this means is that a translocation has a 100 times greater probability of being fixed if chromosome disjunction at anaphase I usually is alternate rather than half alternate and half adjacent. Perhaps this is why novel translocations are relatively common in *Clarkia* and other genera where alternate disjunction is common.

Even though the probability of fixation of a novel variant may be extremely low in a local population, it does not negate the possibility of fixing some novel variant in the species as a whole. If a species contains 10,000 populations, each with one novel chromosomal heterozygote, and the probability of fixing the new arrangement were 1 in 10,000, there would be a very good chance of fixation somewhere.

Note that this is not the end of the story. These populations and their descendants get additional chances to fix chromosomal novelty as long as variation is produced. The species with the highest rates of chromosomal rearrangement would have the highest levels of local establishment per generation or block of consecutive generations, all else being equal. Viewed this way, it is not surprising that species with a penchant for structural change, such as *Clarkia*, would have some chromosomally divergent populations.

CHROMOSOME RACES

The genesis of a new chromosomal race depends on the spread of the novel arrangement beyond the initial site of fixation. The new arrangement may or may not have to replace the existing one. If, for example, the new arrangement was associated with genes that conferred adaptation to a different habitat, the new race could spread unimpeded through the parental system. If no new adaptations were incorporated, the new arrangement would have to replace the existing one in some or all of its range.

Differences between chromosomal races often reflect the incorporation of more than one change. Depending on the overall size of a population or population system and the level of migration between populations, one rearrangement may be fixed locally or globally before the second is increasing, or more than one arrangement may be in transit at the same time.

Aneuploid changes in chromosome number are dependent on at least one reciprocal translocation (Stebbins, 1971). Given our understanding of the dynamics of chromosomal evolution, this process is likely to be exceedingly rare and involve a very large number of generations, unless the level of underdominance is very modest. Correlatively, the divergence of populations for multiple arrangements not affecting chromosome number also must involve a very large number of generations, unless the level of underdominance is very modest.

Races resulting from chromosomal rearrangement have been described in numerous species. In a minority of the cases rearrangements have led to changes in chromosome number. For such to occur, at least two reciprocal translocations are required, as discussed below. Races sharing the same chromosome number are discussed first.

Rearrangement races

Reciprocal translocations have lead to the formation of two widespread and four local chromosome races in the tetraploid *Clarkia rhomboidea* ($2n = 24$; Mosquin, 1964). The most expansive arrangement, Northern, extends from northern California, Oregon, and Washington to Utah and Idaho. The other widespread arrangement, Southern, occurs in southern California and Arizona.

The maximum prophase I configurations of interpopulation hybrids are summarized in figure 2.6. Hybrids between the Northern and Southern race often have nine pairs of chromosomes and a chain of six. This indicates that the races differ by two translocations involving three pairs of chromosomes. Northern differs from Winnemucca by one translocation, and Southern differs from Winnemucca by three translocations involving four chromosomes. The Figueroa and Winnemucca arrangements ostensibly are derived from the Northern type, from which they differ by only one translocation. The Kyburz arrangement apparently is derived from the Southern type, from which it differs by one

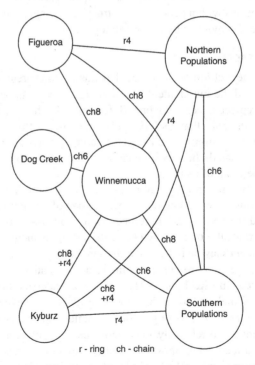

Figure 2.6. Summary of the maximum meiotic metaphase I configurations observed in F_1 hybrids between populations of *Clarkia rhomboidea*. Redrawn from Mosquin (1964), with the permission of the Society for the Study of Evolution.

translocation. Mosquin (1964) concluded that Northern is the original arrangement. Translocations in the species involve only 6 of the 12 chromosomes in the genome.

Paracentric inversions also have played a role in the diversification of *C. rhomboidea,* as indicated by anaphase bridges and fragments. For example, the Southern arrangement differs from Figueroa by at least one inversion and from Winnemucca by at least two.

The congener *C. speciosa* has developed two chromosomal races in California (Bloom and Lewis, 1972). These races, referred to as North and South, differ by a minimum of seven translocations involving eight of the nine chromosomes of the haploid set. Artificial hybrids have a ring of 16 chromosomes and one bivalent at prophase I of meiosis. The races intergrade over a distance of about 30 miles. Bloom and Lewis (1972) analyzed a transect through this zone. Rather than being composites of the parental arrangements, the intergrade populations contain at least 10 novel arrangements, each quite restricted in space (figure 2.7). These arrangements are distributed such that crosses

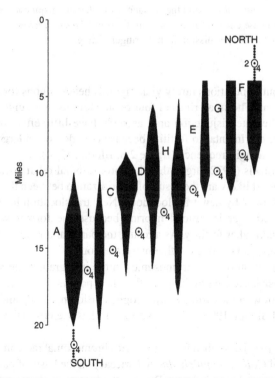

Figure 2.7. Distribution and relative frequency of different chromosome arrangements along the Deer Creek transect in *Clarkia speciosa.* The width of the pattern indicates its approximate frequency. The chromosomal relationships between adjacent arrangements are noted. Adapted from Bloom and Lewis (1972).

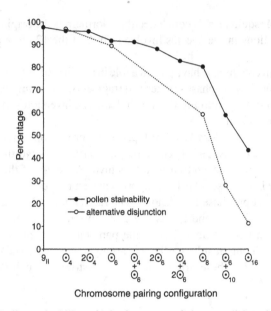

Figure 2.8. Pollen stainability and the frequency of alternate disjunction associated with different prophase chromosomal configurations in *Clarkia speciosa.* Redrawn from Bloom (1974), with permission from Springer-Verlag.

between adjacent populations rarely yield hybrids heterozygous for more than two translocations. The small rings of four or six chromosomes undergo a high frequency of alternate disjunction and, as such, have little effect on fertility. Larger rings are detrimental to fertility because they do not undergo alternate disjunction with a high frequency (figure 2.8; Bloom, 1974).

Bloom and Lewis (1972) argue that the chromosomal relationships among translocations most likely are the result of hybridization between the North and South races followed by new translocations. New translocation heterozygotes would be favored over interracial hybrids because the former would have smaller rings and higher fertility. As a result, the parental arrangements would be replaced in all but the edges of the intergrade zone.

The reorganization of the chromosomes in the intergrade zone reduced an impermeable barrier to interracial gene flow to a porous one. Evidence of interracial gene flow comes from morphological (Bloom, 1976) and allozyme data (Soltis and Bloom, 1991). The primary thrust has been from the South race into the North.

Stoutamire (1977) described four parapatric chromosomal races in the diploid ($n = 17$) annual *Gaillardia pulchella* (Asteraceae) on the basis of pairing irregularities in interpopulation hybrids. Despite the barrier to gene exchange that the chromosomal changes impose, there is little allozyme divergence between the races (Heywood and Levin, 1984). This suggests that the races are of relatively recent vintage. A somewhat similar situation is found in *Helianthus de-*

bilis, where two groups of populations differed from each other by at least two translocations but had not undergone appreciable allozymic divergence (Heiser, 1956; Wain, 1983).

Numerical races

Races involving chromosome number changes have been described in many species. Most, however, involve changes in ploidal level, discussed in chapter 4. Aneuploid races are discussed here. As is the case with rearrangement races, a species may have one major race and one minor race or a more complex assemblage. In some species, rearrangement races have arisen within chromosomal races.

Chromosome evolution in *Ornithogalum tenuifolium* (Hyacinthaceae) has led to the formation four chromosome races ($2n = 4, 6, 8, 12$). Stedje (1989) proposed that the race with $2n = 12$ is the oldest and that we are looking at a descending aneuploid series. This race occurs in tropical east Africa, whereas that with $2n = 4$ occurs in South Africa and Zimbabwe. The other races occur between these two regions. The southward spread of the species may have been accompanied by a reduction in chromosome number through unequal translocations and the loss of centromeres. The pairing configurations observed in hybrids between the $2n = 6$ and $2n = 12$ races indicate multiple translocation differences between them.

Another stepwise progression has been reported by Brighton (1978) in *Crocus cancellatus* (Iridaceae). This species has races with $2n = 8, 10, 14$, and 16. The race with highest chromosome number occurs in Greece and western Turkey, and the one with the lowest number occurs in Lebanon. In general, chromosome numbers decline toward the east. Brighton speculated that this is a descending aneuploid series. Narrowly distributed chromosome races also occur across the same region in *C. speciosus,* but in this species numbers increase toward the east (Brighton et al., 1983). Chromosome races also occur across southeastern Europe in *C. heuffelianus* (Brighton, 1976). The heterogeneity of chromosome numbers within species is not surprising when one considers that closely related species often differ in chromosome number. The penchant for chromosome breakage seems relatively high in *Crocus.*

Consider next some examples of species in which chromosome number races have diversified into population systems with different chromosome arrangements. *Haplopappus gracilis* is one species that has minor numerical and rearrangement races. It is notable because of its unusually low chromosome number ($2n = 4$). A three-paired race ($2n = 6$) occurs in a small area of south-central Arizona (Jackson, 1965). This race probably arose from its widespread relative.

The $2n = 4$ system contains two structurally different races. Whereas one occupies most of the species range in the southwestern United States and Mexico, the other is restricted to the Big Bend area of Texas (United States) and to Chihuahua, Durango, and Sonora in Mexico. The Mexican race is distinguished by a change in centromere position, which probably was due to a centric transposition on the longest of the two chromosomes (Jackson, 1973). This would involve two breaks on either side of the centromere and another break in one arm.

Two numerical races also occur in the self-incompatible, annual composite *Calycadenia pauciflora* (Carr, 1975). A population system with six pairs of chromosomes resides on the outer slopes of the inner North Coast Range of California, whereas another system with five pairs of chromosomes (Pauciflora) resides on the inner slopes of the inner North Coast Range. The six-paired race is divisible into five rearrangement population systems (Elegans, Tehama, Ramulosa, Wurlizer, and Healdsburg) on the basis of abnormal pairing configurations in their synthetic hybrids (Carr, 1975; Carr and Carr, 2000). The races differ by two to four translocations and, in some instances, by inversions.

The single large metacentric chromosome of Pauciflora is homologous with substantial portions of two chromosomes of each of the six-paired races. Thus, Carr (1975) concluded that Pauciflora is the product of chromosome fusion leading to an aneuploid reduction in chromosome number. The six-paired races differ by up to three translocations as evident in the formation of one or more rings and chains in their hybrids. The number of translocations differentiating some races may exceed the number suggested by the meiotic configurations because of redundant translocations involving chromosomes that already have been modified by translocations.

Diakinesis in hybrids between the six-paired Tehama and Wurlizer showing four bivalents and a multiple of four chromosomes is depicted in figure 2.9. More complex pairing between six-paired entities is seen in a PMC from a hybrid between Elegans and Tehama, that contains two bivalents, one branched chain of three and one linear chain of five (figure 2.10). Chromosome pairing in hybrids between five- and six-paired entities may be even more complex. For example, a cell shown in figure 2.11 from a hybrid between Elegans and Pauciflora has one bivalent, one chain of three, and one multiple of six chromosome.

The mean pollen stainabilities of interracial hybrids are depicted in figure 2.12. It is notable that some hybrids between races differing in chromosome numbers have higher pollen viabilities than some hybrids between same-numbered races.

Carr (1981) produced F_2 and F_3 progenies from crosses between Pauciflora ($2n = 10$) and Tehama ($2n = 12$), which differ by three translocations. These progenies contained some plants with the Pauciflora chromosome complement and plants with the Tehama complement in homozygous conditions. Moreover, some plants were recombinants, two of which had a unique chromosome arrangement as revealed by a change in chromosome size and symmetry and by unique meiotic configuration. One rearrangement apparently was a result of a translocation in the F_1 hybrid. The nature of the second was not evident. Intercrossing heterozygotes with the same arrangements should yield a novel true-breeding homozygote. What is important here is that the hybridization of different structurally homozygous races can produce a chromosomal novelty that itself can form the basis of a new race.

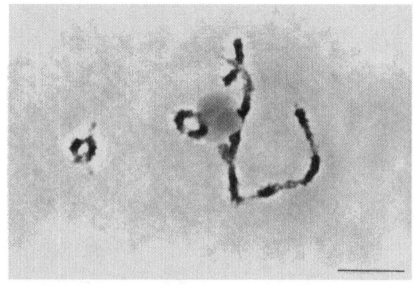

Figure 2.9. Chromosome pairing relationships (4II + 1IV) in a hybrid between two six-paired *Calycadenia* races (Elegans and Tehama). Scale bar = 10mµ. Courtesy of Gerald Carr.

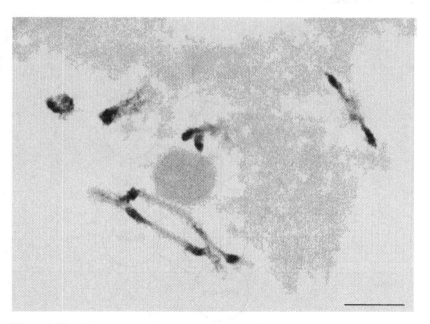

Figure 2.10. Chromosome pairing relationships (2II + 1V + 1III) in a hybrid between two six-paired *Calycadenia* (Tehama and Wurlizer). Scale bar = 10mµ. Courtesy of Gerald Carr.

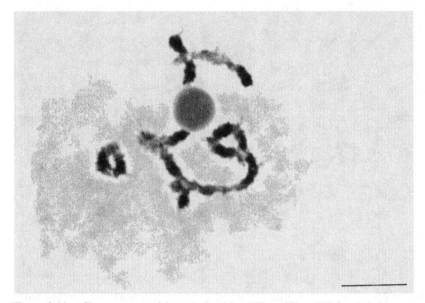

Figure 2.11.   Chromosome pairing relationships (1II + 1VI + 1III) in a hybrid between a five-paired and six-paired *Calycadenia* (Elegans and Pauciflora). Scale bar = 10mµ. Courtesy of Gerald Carr.

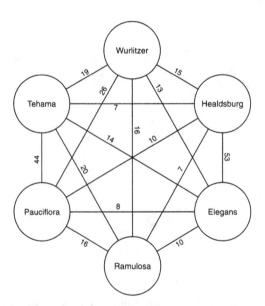

Figure 2.12.   Mean pollen stainabilities (%) of interracial hybrids in *Calycadenia pauciflora*. Redrawn from Carr and Carr (2000) with permission of the Botanical Society of America.

THE SEED BANK AS A SOURCE OF CHROMOSOMAL NOVELTY

Chromosome breakage may be a property of an aging seed bank. In many species, especially annuals and weeds, seeds may spend from a few to many years in the ground before they germinate. As a cohort of seeds ages to the point where less than 50% are viable, genic and chromosomal mutations begin to accumulate (Levin, 1990). Typically there is a significant correlation between visible chromosomal aberrations and the incidence of heritable point mutations. Most genic and chromosomal change is related to lesions in single-stranded DNA and to the failure of their repair before the semiconservative replication that occurs before germination (Osborne, 1982).

The induction of chromosomal aberrations with seed aging was first observed by Navashin (1933) in *Crepis tectorum* (Asteraceae). He found that 80% of plants from 8-year-old seeds had chromosomal chimeras, versus 0.1% in 1-year-old seeds. In the wheat cultivar Baart, 0.1% of 2-year-old seeds had anaphase bridges and fragments in the first mitotic division of root tips, versus 1.5% in 26-year-old seeds and 5.1% in 33 year-old seeds (Gunthardt et al., 1953). In the onion cultivar Yellow Strassburg 1.6% of the first mitotic division in root tip cells of 1- to 2-year-old seeds contained bridges and fragments compared with 3.6% in 3- to 4-year-old seeds and 7.5% in 4- to 5-year-old seeds (Nichols, 1941).

Some of the breaks present in the early stages of development are present in PMCs. Chauhan and Swaminathan (1984) found that 12% of the PMCs in barley from aged seed had chromosomal aberrations versus 20% of the root-tip cells. In soybeans, they found that 7% of the PMCs were abnormal versus 11% of root-tip cells. The presence of chromosomal aberrations in the PMCs is important because it signifies that they can be transmitted to the next generation.

Some inversions and translocations that arose in aged seed indeed are transmitted to the next generation. Gerassimova (1935) reported that in *Crepis tectorum* 10% of the progeny of plants from aged seed had chromosomal rearrangements. These were transmitted through both eggs and sperm. In barley and soybeans, the progeny from aged seed had approximately the same percentage of chromosomal aberrations as did the parental plants (Chauhan and Swaminathan, 1984).

Populations with persistent seed banks are likely to have higher levels of chromosomal polymorphism than populations without such banks, all else being equal. Correlatively, novel rearrangements are more likely to be fixed in populations with persistent seed banks.

OVERVIEW

Chromosomal heterogeneity independent of polyploidy has been observed in a multitude of plant species as judged from analyses of meiotic configurations and to a lesser extent from karyotypic differences. Meiotic configurations offer more insights than do karyotypes into the nature of the chromosomal rearrangements underlying karyotypic change, because the bases for the latter typically cannot be deduced. Whereas cytological approaches often reveal substantive rearrangements, smaller rearrangements undoubtedly go unnoticed. Thus, we may be substantially underestimating the amount of chromosomal variation within species.

Translocations seem to be the most common rearrangement within plants. One finds variation for end arrangements within populations of many species and fixed differences between populations in a smaller number of species. In some instances, fixed differences are accompanied by a change in chromosome number, either ascending or descending. Our understanding of the location of chromosomal break points is rather meager. We need dense genetic mapping of different conspecific populations to truly appreciate the specific character of chromosomal rearrangements.

Translocations are unusually prevalent in the Onagraceae and are well documented in many species of *Clarkia* and *Oenothera*. As discussed in chapter 5, many species of *Oenothera* are obligate chromosomal heterozygotes, in which all of the chromosomes are involved in translocations. The presence of translocation heterogeneity in *Clarkia* in part arises from its penchant for alternate chromosome disjunction at anaphase I in heterozygotes and perhaps heterozygote advantage. The presence of translocation heterogeneity also arises from a penchant for chromosome breakage that may be enhanced by inbreeding.

Despite our substantial knowledge about chromosome races, we know almost nothing about the extent of genetic divergence among them. We would like to know to what extent races have diverged in neutral gene frequencies. We would also like to know whether the imposition of a chromosomal barrier to interpopulation gene exchange has been followed by adaptive divergence.

Aging seed banks may be an important source of chromosomal rearrangements for species that live in unpredictable habitats. Unfortunately, our understanding of this matter is largely restricted to crop plants whose seed ages are artificially accelerated.

The fixation of chromosomal rearrangements occurs in the face of the heterozygote being partially sterile. Accordingly, unless there is some form of heterozygote advantage, it is likely to occur as a result of stochastic processes, and these are most pronounced in populations with small effective sizes. Self-fertility promotes fixation because self-fertilization depresses the effective population size. Novel chromosome rearrangements are more likely to be fixed in populations where heterozygotes have a tendency to undergo alternate disjunction at anaphase I, all else being equal. Chromosomal novelty is unlikely to diffuse through an established population system, because each time it is introduced into a population it will be at a gross disadvantage. Novel chromosome races are most likely to be the products of local fixation followed by geographical expansion rather than replacement.

3

Euploid and Aneuploid Diversification within Genera

Chromosomal diversification within genera has been a focal point of plant evolutionary studies. There are two prime reasons for this interest. First, chromosomal change imparts a partial or complete barrier to interspecific gene exchange. Second, chromosomal traits may provide clues to the relationships of species.

A considerable amount has been written about the nature of chromosomal change. This chapter deals with alterations in the chromosome complement that are due to chromosomal rearrangements or an increase or decrease in genome size. Changes in chromosome numbers due to rearrangements are discussed below, whereas those related to polyploidy are discussed in chapter 6.

KARYOTYPE EVOLUTION

Many chromosomal studies of genera include the description of species' karyotypes. The karyotype is the phenotypic aspect of the chromosome complement as seen at mitotic metaphase. For several decades, the karyotype was the prime cytological trait used to infer species relationships.

A description of the karyotype of a species includes the following attributes: (1) the chromosome number, (2) the total length of the chromosome complement (genome size), (3) the absolute and relative sizes of chromosomes, (4) the symmetry of each chromosome as dictated by the position of the centromere on each chromosome, (5) the number and positions of satellites associated with the nucleolar-organizing regions, and (6) the distribution of heterochromatic segments.

Differences in chromosome number may arise through multiple translocations. Stebbins (1950) described progressions leading to an increase and decrease of centromeres without a change in genome size through two translocations arising in single plants followed by the production of $n + 1$ and $n - 1$ gametes (figure 3.1). In contrast to Stebbins's model, Schubert and Rieger (1985) propose that new numbers may arise from the crossing of two plants each with a unique translocation followed by the production of $n + 1$ and $n - 1$ gametes.

Robertsonian fusions and fissions also may alter chromosome number (Jones, 1998). Robertsonian fusion is a process wherein two nonhomologous acrocentric chromosomes each break at the centromere. This is followed by the fusion of the long chromosome arms and the fusion of the short chromosome arms. The result is a long metacentric chromosome and a short metacentric chromosome. Fusion of gametes carrying both chromosomal novelties produces a homozygote with the original chromosome number but altered karyotype. Frequently, the fusion chromosome incorporating the short arms of the broken chromosomes is lost, thus yielding a homozygote whose base number is one less than the original. However, the total number of long arms remains the same. Progressive fusions can produce a steady decline in chromosome number.

Fissions involve chromosome breakage at the centromere. They are most likely to occur in chromosomes with two relatively long arms. Each fission will increase the base chromosome number by one. Fissions are associated with heterochromatin-rich metacentric chromosomes, whereas fusions tend to occur preferentially between chromosomes with less heterochromatin (Kenton et al., 1993). Fission and fusion have been especially active in *Cymbispatha* (formerly a section of *Tradescantia;* Commelinaceae).

Alterations in chromosome symmetry may arise through translocations, pericentric inversions (the centromere is included in the inverted segment), or fusion. Genome size differences may arise from the addition or subtraction of redundant DNA, as discussed in chapter 1.

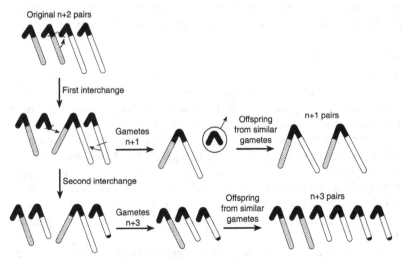

Figure 3.1. The increase and decrease of chromosome number through reciprocal translocations. Regions adjacent to the centromere are shown in black. Shaded chromosome arms are homologous, as are white arms. Redrawn from Stebbins (1950) with permission of Columbia University Press.

Karyotype diversity

Most genera of herbaceous angiosperms display interspecific differences in chromosome size and symmetry, if not number. One well-known genus with quite disparate karyotypes is *Crepis,* whose chromosomes were studied by Babcock (1947), Jones and Brown (1976), and most recently by Dimitrova and Greilhuber (2000). The species differ in base number (from three to six), chromosome size, and symmetry. The karyotypes of eight Bulgarian species and their 1C DNA contents based on the findings of the latter investigators are shown in figure 3.2.

Brachyscome is a genus with even more divergent diploid karyotypes (Watanabe et al., 1999). The species differ in base number from two to nine, as well as in chromosome size and symmetry. The degree of variation among

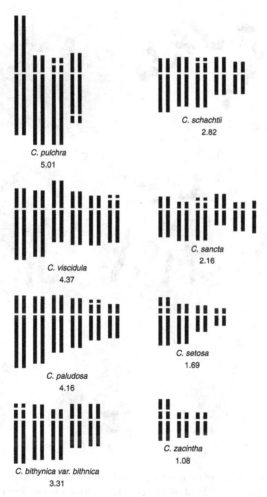

Figure 3.2. Idiograms of the chromosomes of eight Bulgarian *Crepis* species and their 1*C* DNA contents. Redrawn from Dimitrova and Greilhuber (2000), with permission of Academic Press.

species is most unusual. Total chromosome length varies from 25 to 82 μm. The mean level of chromosome asymmetry varies from 0.09 to 0.50. A value of zero indicates perfect symmetry; that is, both chromosome arms have the same length. The value increases as the centromere moves toward one end of the chromosome, reaching its maximum near 0.5 when centromeres are near the ends of the chromosomes. The chromosomes of four species are shown in figure 3.3.

One aspect of the karyotype is not necessarily linked to another, as illustrated using chromosome size in relation to two other traits. For example, species that are quite divergent in chromosome symmetry may differ little in total chromosome length. *Nicotiana tomentosiformis* and *N. tomentosa* have chromosomes of similar total lengths, yet *N. tomentosiformis* has seven chromosomes with me-

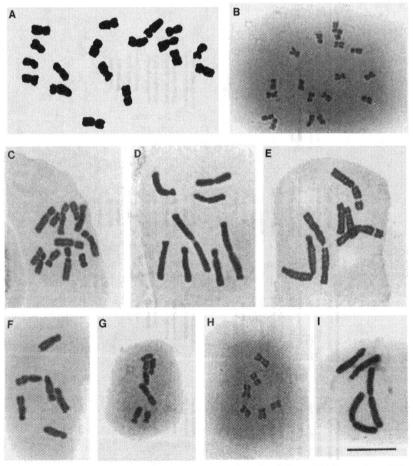

Figure 3.3. Chromosomes of (a) *Brachyscome decipiens* (2*n* = 18), (b) *B. halophila* (2*n* = 18), (c) *B. rara* (2*n* = 12), (d) *B. eriogona* (2*n* = 8), (e) *B. muelleri* (2*n* = 6), (f) *B. debilis* (2*n* = 6), (g) *B. smithwhitei* (2*n* = 6), (h) and (i) *B. dichromosomatica* (2*n* = 4). Scale bar = 10mμ. Courtesy of Kuniaki Watanabe.

dian centromeres and five chromosomes with subterminal centromeres, whereas
N. tomentosa has three chromosomes with median centromeres, four with sub-
median centromeres, and five with subterminal centromeres.

There is no mandatory relationship between DNA content and chromosome
number. For example, in *Crepis* species with $x = 4$ have the same genome sizes
as species with $x = 5$ (Jones and Brown, 1976). In *Pennisetum*, species with the
base number of $x = 5$ and $x = 9$ have less DNA per genome than species with
$x = 7$ (Martel et al., 1997).

The power of karyotype analyses expanded in the 1960s with the development of Giemsa staining. This procedure reveals heterochromatic C-bands near
the centromere and sometimes in interstitial and terminal regions as well. The
genus *Aegilops* (Poaceae) provides a striking example of their utility in differentiating species' karyotypes (Badaeva et al., 1996a). The C-banding patterns
of four species ($n = 7$) are shown in figure 3.4. In the absence of C-bands, it
would be difficult to distinguish the karyotypes of some species from others.

Giemsa staining is only one procedure to elicit chromosome banding in
plant chromosomes. Other procedures and examples of their applications are
reviewed by Fukuda (1984).

The 1980s brought an even more powerful tool: genomic *in situ* hybridization (GISH). With *in situ* hybridization, one can physically map very specific regions of the genome such as the 5S and 18S–26S ribosomal RNA gene families.
Even closely related species may differ in the spatial organization of these gene
families. For example, diploid species of *Aegilops* ($n = 7$) differ in the number
and positions of 5S and 18S–25S sites (Badaeva et al., 1996b). The 18S–25S
loci are located on homoeologous (partially homologous chromosomes of related species) chromosome groups 1, 5, and 6 (figure 3.4). One or two 5S loci
occur on the short arms of chromosome groups 1 and 5. Diploid species of
Hordeum ($n = 7$) also differ in the number, location and relative order of 5S
and 18S–25S rDNA sites (Taketa et al., 1999). The number of 18S–26S sites
varies from 4 to 20, one or two of which are major. In both genera the phylogenetic conclusions based on these sites agree with other types of data regarding species relationships and the origin of polyploid taxa.

Karyotype trends

Given that the karyotypes of species in hundreds of genera have been described,
it is important to know whether there are patterns of karyotype evolution within
genera. Stebbins (1971, 1976) provides evidence that there are patterns, although
they are not universal. The first pattern involves genome size. In some genera,
increasing specialization has been accompanied by a reduction in genome size,
especially when the specialization involves a shift from perennial to annual habit
or to a shorter growing season. Genera purported to have descending genome
sizes include *Crepis* (Jones and Brown, 1976), *Myosotis* (Grau, 1964), *Artemisia*
(Nagl, 1974), *Asphodelus* (Asphodelaceae; Diaz Lifante, 1996), *Allium* (Ohri
et al., 1998), *Helianthus* (Sims and Price, 1985), *Ranunculus* (Smith and Bennett, 1975), *Podolepis* (Konishi et al., 2000), and *Arachis* (Singh et al., 1996).

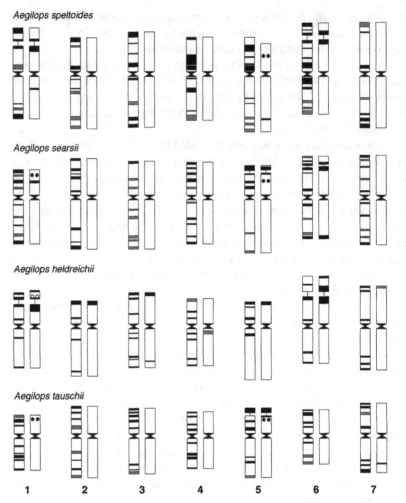

Figure 3.4. Idiograms of the seven chromosomes of *Aegilops speltoides, Ae. searsii, Ae. heldreichii,* and *Ae. tauschii* after C-banding (left) showing locations of 5S (circles) and 18S–26S (rectangles) ribosomal RNA loci (right). Redrawn from Badaeva et al. (1996a,b), with permission of the National Research Council of Canada.

Conversely, increasing genome size ostensibly has accompanied the shift to the annual habit or shorter growing season in *Anthemis* (Nagl, 1974), *Chrysanthemum* (Dowrick and El-Bayoumi, 1969), and *Lolium* (Rees and Jones, 1967). In *Anacyclus* (Nagl and Ehrendorfer, 1974; Nagl, 1974), *Plantago* (McCullagh, 1934; Rahn, 1957), and *Vicia* (Chooi, 1971) both increasing and decreasing trends occur.

In some genera, ecological shifts also have been associated with changes in chromosome numbers. For example, the shift from a perennial to annual habit in arid landscapes appears to have been accompanied by a reduction in chromosome number in *Crepis* (Babcock, 1947), *Calotis* (Stace, 1978, 1982), *Podolepis* (Konishi et al., 2000), and *Haplopappus* (Jackson, 1962; Smith, 1966).

Karyotypic trends also involve chromosome symmetry. This was first noted by Levitsky (1931a,b) based on the studies in the Helleboreae (Ranunculaceae). He believed that, in this and other groups, chromosomes became more asymmetrical as evolution progressed. This idea was embraced by Stebbins (1950, 1971), who contended that in the Asteraceae the development of asymmetric karyotypes is associated with entrance into pioneer habitats and often with a shift from the perennial to annual habit.

Stebbins (1971) proposed that increasing asymmetry might be achieved through natural selection for an accumulation of adaptive clusters of linked genes on one chromosome arm, which would tend to lengthen that arm at the expense of the other. I propose an alternative hypothesis for increasing asymmetry when it is associated with a decline in genome size. In many genera, contraction of genome size is achieved by an equal reduction in DNA per chromosome regardless of chromosome size (*Vicia,* Raina and Rees, 1983; *Papaver,* Srivastava and Lavania, 1991). If the same amount of DNA was removed from both chromosomes arms, asymmetrical chromosomes would become more asymmetrical. A correlate to this would be that increasing genome size would be accompanied by greater chromosome symmetry. Unequal reciprocal translocations, fusions, and fissions would disrupt the expected pattern that is based solely on genome size alterations. The aforementioned interpretations of karyotype change with evolutionary advancement are not based on a molecular phylogeny. This is an obvious problem, because we usually do not have strong insights into which species are basal. Fortunately, molecular phylogenies have been used in conjunction with some karyotypic analyses.

Hypochaeris (Asteraceae) is one genus whose karyotypic attributes were mapped onto a molecular phylogeny. In this case, the phylogeny was based on sequences of the internal transcribed spacers (ITSs) of the 18S–26S nuclear ribosomal DNA (Cerbah et al., 1998a,b). The species differ in base number ($x = 3, 4, 5,$ and 6), genome size, and chromosome symmetry. There was both an increase and decrease in chromosome number and in DNA content within the genus. It is noteworthy that the level of chromosome symmetry is positively correlated with DNA content. This is what would be expected if the addition and reduction of DNA content were equal in each chromosome arm.

In contrast to *Hypochaeris,* karyotype evolution in the closely related slipper orchids *Phragmipedium* and *Paphiopedilum* appears to have been unidirectional (Cox et al., 1998). Chromosome numbers vary from $2n = 18$ (all metacentric chromosomes) to $2n = 30$ in the former. Yet the number of chromosome arms is largely conserved. The phylogenetic tree indicates that a diploid karyotype of 18 metacentric chromosomes is plesiomorphic (basal), and that additional chromosomes with a telocentric form were generated through centric fission. Chromosome numbers in *Paphiopedilum* range from $2n = 26$ to $2n = 42$. Nevertheless, the number of chromosome arms in this genus is almost completely conserved. In *Phragmipedium,* centric fission of metacentric chromosomes into telocentrics is the predominant mechanism of karyotype evolution. The plesiomorphic karyotype for *Paphiopedilum* ostensibly was composed of 26 metacentric chromosomes.

Karyotype analysis has one serious deficiency. Polyploidy aside, one rarely can determine the chromosomal changes that are responsible for divergent karyotypes because of the paucity of chromosome markers. This problem may be overcome in related taxa where there are dense genetic maps, because genes rather than heterochromatic bands or knobs will indicate the identity of chromosome arms or segments thereof. In the absence of such maps, the best we can do is to study chromosome pairing in hybrids between karyotypically divergent plants with the hope of observing configurations indicative of translocations, inversions, fusions, and fissions and possibly identifying the chromosomes involved.

Finally, whereas karyotypic differences among species are likely to be accompanied by abnormal pairing of their chromosomes in F_1 hybrids, we should not assume that pairing will be normal in species with very similar or essentially identical karyotypes. Consider the sibling species *Gibasis karwinskyana* and *G. consobrina.* Their karyotypes are almost identical with Feulgen staining, and their genome sizes are similar (Jones et al., 1975). However, their chromosomes fail to pair in F_1 hybrids. The chromosome complements of each species are readily identifiable in hybrids using GISH (Parokonny et al., 1992). This signifies substantial molecular divergence between the chromosome complements. This is manifested in the low level of duplex formation from the species' single-stranded DNAs.

CHROMOSOMAL EVOLUTION AMONG CONGENERIC SPECIES

Molecular cytogenetics

High-density genetic maps have become a potent tool for studying chromosome evolution. By comparing genetic linkage maps based on a common set of markers, one can directly identify homoeologous loci and collinear chromosomal segments. This information may shed light on ancestral chromosomal rearrangements and on evolutionary relationships between different chromosomes.

The most informative data set on the chromosomal architecture of diploid wild plants is for *Helianthus annuus* and *H. petiolaris.* Rieseberg, VanFossen, and Derochers (1995) generated linkage maps based on 212 loci in *H. annuus* and 400 loci in *H. petiolaris.* The genetic markers were mapped to 17 linkage groups corresponding to the haploid chromosome number of the species. By comparing the linear order of homologous markers and their genomic locations, Rieseberg et al. were able to infer chromosomal structural relationships between the species. *H. petiolaris* and *H. annuus* differ by at least 10 separate structural rearrangements, which include at least seven reciprocal translocations and three inversions. First-generation interspecific hybrids are semisterile. Seed viability is less than 1% and pollen viability is less than 10%.

Substantial chromosomal change also has accompanied the evolution of the legumes *Vigna radiata* (mungbean) and *V. unguiculata* (cowpea) from a common ancestor. Both species have the same chromosome number ($n = 11$). Menancio-Hautea et al. (1993) mapped 171 loci distributed on 14 linkage groups for mungbean and 97 loci on 10 linkage groups for cowpea. They compared the two genomes on the basis of the linkage arrangement of 53 shared markers. Several large linkage groups were conserved; others were reorganized (figure 3.5). Regarding the latter, mungbean linkage group 1 is distributed on cowpea groups 2 and 3. Mungbean group 4 is located on cowpea groups 4, 5, and 6. These changes were the result of several translocations. The linear order

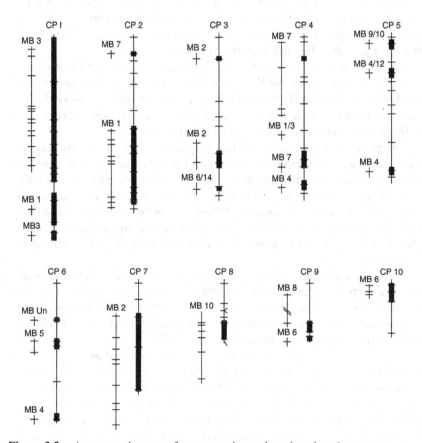

Figure 3.5.   A comparative map of cowpea and mungbean based on the cowpea linkage groups. Cowpea linkage groups are indicated by CP at the top of each figure; mungbean linkage groups are designated by MB followed by the corresponding linkage group number. Conserved regions between the species are indicated by black bars in the cowpea linkage groups. The corresponding segment in the mungbean linkage groups are shown to the left of the cowpea figure. Redrawn from Menancio-Hautea et al. (1993), with permission of Springer-Verlag.

of loci was altered within one conserved linkage. Much of the conserved group on cowpea group 1 is inverted with respect to that from mungbean.

Inversions may play a prominent role in reorganizing the chromosomal architectures of congeneric species as in *Solanum*. Potato (*S. tuberosum*) and tomato (*S. esculentum*) differ by five chromosomal rearrangements, all of which involve breaks at or near the centromere (Bonierbale et al., 1988; Tanksley et al., 1992). These rearrangements result in paracentric inversions of the short arms of chromosomes 5, 9, 11, and 12 and of the long arm of chromosome 10 on the tomato map. The inversions collectively constitute 11% of the genome.

Potato and tomato share the A genome, one of five basic genomes in the section *Pepota* as judged from chromosome pairing and fertility relationships (Hawkes, 1990). Because species hybrids from intergenomic crosses have irregular meiosis in contrast to fairly regular meiosis in intragenomic hybrids, we would expect species with different genomes to be more divergent than those that share genomes. This indeed is what we find.

Comparative maps also are available for *Solanum palustre* and *S. etuberosum*. These species share the E genome. Perez et al. (1999) compared the A and E genomes and found that most linkage groups are largely conserved. The genomes differ in linkage group 9 by inversions and a transposition. They also differ in translocations involving groups 1, 2, 4, 8, and 12.

The genomes of species in different genera may differ more than species within genera. For example, the genomes of tomato and pepper (*Capsicum annuum*) differ by 30 chromosome breaks since divergence from a common ancestor (Livingstone et al., 1999) versus 10 since the divergence of tomato and potato from a common ancestor. Whereas all of the breaks involved in the latter yielded inversions, the breaks involved in the tomato–pepper divergence yielded at least 5 translocations, 10 paracentric inversions, and 2 pericentric inversions.

Genetic mapping also has provided detailed information on chromosomal differences in *Secale* and *Aegilops S. cereale* differs from *Aegilops tauschii* by seven translocations (Devos et al., 1993), whereas the latter differs from *Ae. longissima* by one translocation (Devos et al., 2000) and does not differ at all from *Ae. speltoides* (Maestra and Naranjo, 1998).

Although species within the same genus have fewer rearrangements than species in related genera, this is not always the case. The genome of barley (*Hordeum vulgare*) is highly collinear with that of *Aegilops tauschii,* apparently differing by only two inversions (Dubcovsky et al., 1996). However, the genomes of *Ae. tauschii* and *Ae. umbelluta* differ by a minimum of 11 rearrangements (Zhang et al., 1998). One cannot explain this counterintuitive result in terms of incorrect taxonomy, because phylogenetic data show that *Ae. umbellulata* is more closely related to *Ae. tauschii* than to barley (Kellogg et al., 1996).

It may be that some genomes fix rearrangements more readily that others. This would explain why *Ae. umbellulata* is so much more deviant from *Ae. tauschii* than are *Ae. longissima,* and *Ae. speltoides*. Uneven rates of fixation also would explain the fact that the genomes of *Sorghum* and *Saccharum* only differ by one translocation and two inversions, whereas *Saccharum spontaneum* and *S. offic-inarum* differ by 11 rearrangements (Ming et al., 1998). A propensity toward rearrangement also may be a property of the pepper genome, because the number of differences between pepper species is equal to the number of differences between peppers and tomatoes (Tanksley et al., 1992). A propensity for rearrangement in pearl millet seems to explain why this species and foxtail millet genomes are highly rearranged, whereas the genomes of millet and rice are less so (Devos and Gale, 2000).

Finally, we may address the issue of rearrangement number as a function of divergence time as judged from comparative linkage mapping. In general, the rearrangement number increases with divergence time (Lagercrantz, 1998). This relationship has been established using intra- and intergeneric comparisons focusing on *Brassica, Zea, Lycopersicon,* and *Gossypium* (figure 3.6). The outlier shown in figure 3.7 involves *Brassica* and *Arabidopsis.* The unusual structural divergence between these genomes may be due in part to the fact that the latter is a degenerate hexaploid, as discussed in chapter 5.

Chromosomal cytogenetics

The organization of chromosome complements within genera have been studied extensively through the analysis of meiotic pairing configurations in F_1 hybrids. This approach has two weaknesses relative to the comparative molecular cytogenetic approach. First, interspecific hybrids must be produced. Because all members of a genus are unlikely to be crossable or produce viable hybrids with normal reproductive structures, meiotic analyses in most genera have been

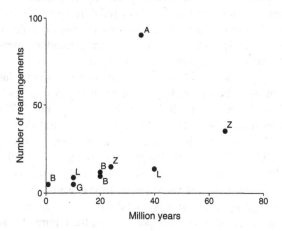

Figure 3.6.   Estimated number of chromosomal rearrangements differentiating species of *Brassica* (B), *Gossypium* (G), *Zea* and *Sorghum* (Z), *Brassica* and *Arabidopsis* (A), and *Lycopersicon* and *Solanum* (L) in relation to divergence times. Redrawn from Lagercrantz (1998), with permission of the Genetics Society of America.

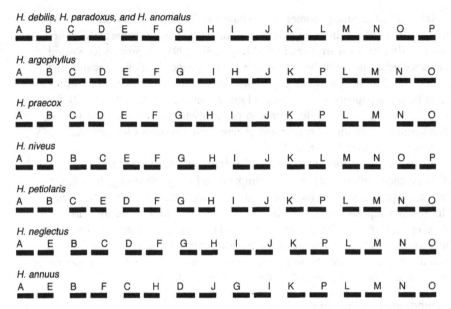

Figure 3.7. Chromosomal end arrangements in *Helianthus*. In addition to the chromosomes shown, each species has nine additional chromosomes with the same end arrangements. Based on Chandler et al. (1986).

restricted to closely related species. This stricture severely limits the scope of one's analyses.

Another problem with meiotic data in species hybrids is that fine-scale rearrangements may be undetected. Only substantial differences are likely to have meiotic signatures as in the case of translocations (rings or chains of four or more chromosomes) and paracentric inversions (bridges and fragments). Small rearrangements may cause chromosomes not to pair.

Some pairs of species produce hybrids with normal pairing relationships. These have been described in many genera including *Agropyron* (Stebbins and Pun, 1953), *Glandularia* (Solbrig, 1968), *Oryza,* (Chu, 1972), *Nigella* (Strid, 1970), *Geum* (Gajewski, 1959), *Primula* (Valentine, 1952), and *Linum* (Seetharam, 1972). In other pairs of species, chromosomes fail to pair as in *Populus* (Peto, 1938), *Gilia* (Grant, 1952), *Geum* (Gajewski, 1959), *Triticum* (Wagenaar, 1970), *Clarkia* (Abdel-Hameed, 1971), *Lycopersicon* (Menzel, 1962), and *Gossypium* (Menzel and Brown, 1955). We cannot say that the species in the first group lack chromosomal differences, nor can we say what kinds of rearrangements (if any) are responsible for the failure of chromosome pairing in the second group.

Chromosomal rearrangements may or may not alter chromosome numbers. Here I discuss examples of chromosomal change that does not affect chromosome number and then chromosomal change that does affect numbers.

Chromosomal change not affecting chromosome number

Where chromosomal diversity exists, the level varies widely among genera. A low level of chromosomal diversity was described by Seavey and Raven (1977a,b) in *Epilobium* sect. *Epilobium* (Onagraceae). All species have the somatic chromosome number of $2n = 36$. From experimental hybrids they recognized four chromosomal arrangements, A, B, C, and D. The B arrangement is the most widespread, occurring in Eurasian, African, and Australasian species, and to a less extent in American species. The A arrangement differs from the B by one reciprocal translocation and occurs in American and European species. The C arrangement, which characterizes a group of circumboreal species, differs from A by two reciprocal translocations and from B by one translocation. Only one species has the D arrangement, which differs from the B by one reciprocal translocation and from A and C by two translocations.

A greater level of chromosomal evolution has occurred among annual diploid species of *Helianthus* ($x = 17$), as described by Chandler, Jan, and Beard (1986). Chromosome pairing relationships indicate that species differ from each other by zero to six translocations and zero to eight paracentric inversions. Based on the observed multivalent configurations in numerous combinations of interspecific hybrids, Chandler, Jan, and Beard proposed end arrangements for nine species, as shown for eight chromosomes in figure 3.7. The species do not differ in their additional nine chromosomes.

It is necessary to assign an ancestral arrangement to understand the directionality in changes of chromosomal end arrangements. The arrangement of *Helianthus debilis* was chosen by the aforementioned researchers because it is shared with some perennials and other annuals. Phylogenetic relationships inferred from sequence variation in the ITS region of rDNA (Schilling et al., 1998) support this contention. Phylogenetic relationships inferred from nuclear rDNA restriction site and length variation (Rieseberg, 1991) suggest that either *H. argophyllus* or *H. annuus* is basal. Both phylogenies indicate that *H. debilis* and *H. praecox* are closely related, as are *H. annuus* and *H. argophyllus* on the one hand, and *H. niveus, H. petiolaris,* and *H. neglectus* on the other.

Referring back to figure 3.7, we see that *H. debilis* and *H. praecox* differ in three end arrangements. The chromosome complements of *H. annuus* and *H. argophyllus* differ in five end arrangements, and *H. petiolaris* and *H. niveus* differ in six end arrangements. Thus, the chromosome complements of even closely related species are well differentiated. If the chromosomal architecture of *H. debilis* were basal, the chromosomal evolution of the other species would have involved changes in as many as eight end arrangements out of the 17 in each species.

The genus *Gaura* (Onagraceae) is notable for the level of chromosomal rearrangement that accompanied its evolution. Carr et al. (1988) compared the chromosome complements of *G. hexandra, G. brachycarpa,* and *G. suffulta* (sect. *Pterogaura*). Hybrids between *G. hexandra* and *G. suffulta* have rings of four chromosomes plus rings of six chromosomes and two bivalents. Hybrids between *G. hexandra* and *G. brachycarpa* have a ring of 12 chromosomes and a bivalent. The species share some end arrangements. Species in the section *Stipogaura* also differ from each other by multiple translocations (Carr et al., 1986b).

It is particularly interesting to consider the work of Carr and Kyhos (1986) on chromosomal diversification in the Hawaiian silverswords, because a molecular phylogeny (Baldwin, 1997) provides guidance on the direction of change. Moreover, chromosomal rearrangement has not been the hallmark of speciation on oceanic islands (Stuessy and Crawford, 1998). The genus *Dubautia* contains five distinctive genomes that differ among each other by at least one reciprocal translocation. Some genomes are shared by two or more species. The species with each of these genomes are as follows: genome I—*D. knudsenii, D. laxa, D. microcephala, D. plantaginea;* genome II—*D. paleata;* genome III—*D. laevigata;* genome IV—*D. scabra, D. latifolia;* genome V—*D. arborea, D. cilio-*

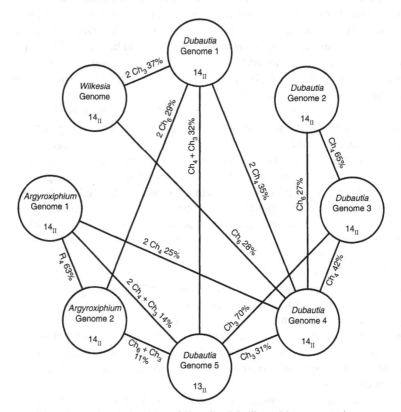

Figure 3.8. Genomic relationships of Hawaiian Madiinae. Lines connecting genomes indicate maximum meiotic chromosome pairing configurations (bivalents omitted) and pollen stainabilities of intergenomic hybrids. Ch refers to chains of chromosomes; R, to rings. Redrawn from Carr and Kyhos (1986), with permission of the Society for the Study of Evolution.

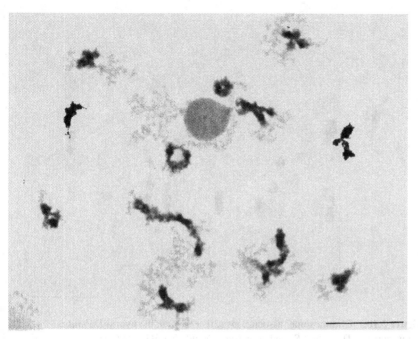

Figure 3.9. Chromosomal associations (12II + 1III) in the hybrid *Dubautia ciliolata* ×
D. scabra. Scale bar = 10mμ. Courtesy of Gerald Carr.

*lata, D. dolosa, D. linearis, D. menziesii, D. platyphylla, D. reticulata, D. shers-
fiana*. All species with genomes I–IV have a somatic number of $2n = 28$. All
species with genome V have a somatic number of $2n = 26$. Based on Baldwin's
18S–26S nuclear rDNA phylogeny, genomes II and III are basal, genome I
is derived from genome III, and genome IV is derived from genome V.

The chromosome pairing anomalies of hybrids between species in the
different genomic assemblages are depicted in figure 3.8. Some species combinations yield hybrids with chains of three or four chromosomes, indicating
heterozygosity for single translocations. For example, hybrids of *Dubautia
ciliolata* (genome V) and *D. scabra* (genome IV) have 12 pairs and a chain
of three chromosomes during diakinesis (figure 3.9). Other combinations yield
hybrids with two multivalents or a chain of six chromosomes, indicating heterozygosity for two translocations. Evidence of two translocations is seen at diakinesis in hybrids between *D. plantaginea* (genome I) and *D. scabra* (genome
IV; figure 3.10).

Carr and Kyhos (1986) found that the two other silversword genera, *Wilke-
sia* and *Argyroxiphium*, are not highly divergent from *Dubautia*. Genomes I and
II of *Argyroxiphium* differ from genome V of *Dubautia* by three translocations.
Genome I of the former and the genome of *Wilkesia* differ from genome IV of
Dubautia by two translocations (figure 3.8).

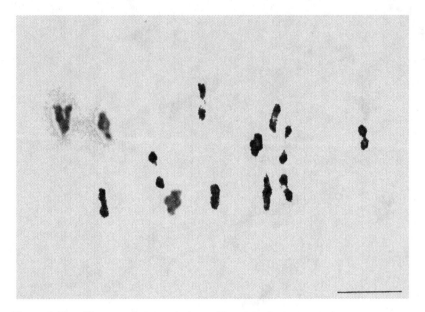

Figure 3.10. Chromosomal associations (2II + 2IV) in the hybrid *Dubautia plantaginea* × *D. scabra*. Scale bar = 10mµ. Courtesy of Gerald Carr.

Based on the examples of chromosomal rearrangement considered thus far, it might be tempting to conclude that the penchant for chromosomal rearrangement within a genus is similar across phylads. This conclusion is not warranted. Consider, for example, the genus *Calycadenia* (Compositae). Carr (1977) crossed species within two natural groups. One group contains *C. multiglandulosa, C. ciliosa, C. pauciflora, C. hispida, C. spicata,* and *C. oppositifolia.* The second group contains *C. villosa, C. mollis, C. truncata,* and *C. tenella.* Hybrids in the first groups are characterized by multivalents, whereas those in the second group are characterized by univalents. We see this in the hybrid *C. mollis* ($n = 7$) × *C. tenella* ($n = 9$), which displays seven large univalents from the former and nine small univalents from the latter (figure 3.11). Hybrids in the first group of species are partially fertile, while those in the second group are almost completely sterile.

Although there are many examples of chromosomal change accompanying speciation, it is important to recognize that chromosomal divergence is not a necessary correlate of morphological divergence and long periods of geographical isolation. For example, paleobotanical and molecular evidence indicates that *Liriodendron tulipifera* of the eastern United States and *L. chinense* of mainland China diverged between 10 and 16 million years ago, but their hybrids exhibit normal meiosis (Parks and Wendel, 1990). Other examples of chromosomal stability among North American and Asian congeners have been described

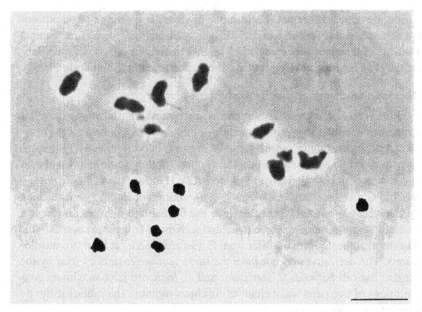

Figure 3.11. Metaphase I with 16 univalents in a hybrid between *Calycadenia mollis* and *C. tenella*. Seven large univalents are from the former and nine small bivalents are from the latter. Scale bar = 10mμ. Courtesy of Gerald Carr.

in *Campsis* (Sax, 1933) and *Catalpa* (Smith, 1941). *Platanus occidentalis* from eastern North America and *P. orientalis* from the eastern Mediterranean probably have been isolated more than 25 million years, and yet their hybrids have normal chromosome pairing and high fertility (Sax, 1933). Indeed chromosomal stasis seems to be a general feature of woody perennials (Stebbins, 1958; Grant, 1981), even in genera radiating extensively on island archipelagos such as *Bidens, Cyrtandra,* and *Lipochaeta* (Carr, 1998).

The bases for chromosomal stasis in some groups are not understood. Perhaps the chromosomes of some groups are less prone to breakage than the chromosomes of others. Perhaps some groups have been subject to fewer genetic bottlenecks, which favor the fixation of novel arrangements. Perhaps stabilizing selection eliminates chromosomal variants in some groups. Kyhos and Carr (1994) have argued the latter on the basis that chromosomal rearrangements may alter the relative positions of genes within the chromosomal complement and that gene function may be negatively affected by such alterations.

If position effect is an important issue, then we would expect congeneric annuals and perennials to have similar levels of chromosomal evolution. However, there are some notable examples where this is not the case. One is in the genus *Chaenactis,* where crosses between morphologically divergent perennial species from remote geographic regions yield hybrids with normal chromosomal pairing in contrast to annual species none of which share the same chromosomal arrangement (Kyhos and Carr, 1994). Other genera where chromosomal stasis in perennials is accompanied by lability in annuals include *Crepis* (Babcock, 1947), *Helianthus* (Heiser et al., 1962), and *Knautia* (Ehrendorfer, 1962).

Chromosomal change affecting chromosome number

In addition to altering end arrangements, reciprocal translocations can alter chromosome numbers. A prime example of this involves *Chaenactis glabriuscula* (Asteraceae), which ostensibly gave rise to *C. stevioides* and *C. fremontii* (Kyhos, 1965). The parental species has a haploid chromosome of $n = 6$ in contrast to $n = 5$ in both derivatives (figure 3.12). Hybrids between *C. glabriuscula* and *C. fremontii* have a maximum configuration of four pairs and a chain of three chromosomes. Hybrids between the former and *C. stevioides* have a maximum configuration of three pairs and a chain of five chromosomes. This indicates that the derivatives differ from *C. glabriuscula* by at least two translocations. The derivatives do not have the same rearrangements as seen by the fact that hybrids between *C. stevioides* and *C. fremontii* have maximum configurations of two pairs and a chain of six chromosomes. The species differ by at least two translocations.

Some of the chromosomes of the three species have distinctive characteristics that made it possible for Kyhos to recognize chromosomes in interspecific hybrids and to analyze the chromosomal arrangements of the species. A diagram of their probable structural arrangements is presented in figure 3.10.

C. fremontii is unique in having its nucleolar organizer on chromosome C instead of chromosome B as in *C. glabriuscula*. A reciprocal translocation is

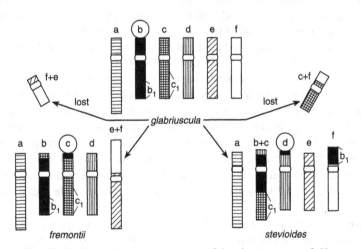

Figure 3.12.   Probable structural arrangements of the chromosomes of *Chaenactis glabriuscula* and its derivatives. Redrawn from Kyhos (1965), with permission of the Society for the Study of Evolution.

the most likely explanation for the transposition. An unequal reciprocal translocation between chromosomes E and F and the loss of a centromere ostensibly explains the reduced chromosome number in *Chaenactis fremontii* ($n = 5$ instead of $n = 6$ in the progenitor) and the genesis of the longest chromosome.

C. stevioides also differs from *C. glabriuscula* in its novel nucleolar organizer position that is on chromosome D. This apparently is the result of a translocation with chromosome B that also has undergone a translocation with chromosome C. Another translocation occurred between chromosomes B and F. These changes were accompanied by the loss of a small chromosome with the centric portion of the C chromosome and the short arm of F, thus reducing the chromosome number of *C. stevioides* to $n = 5$.

The putative direction of evolution in the genus is based on the fact that *C. stevioides* and *C. fremontii* are the only members of the genus with $n = 5$; most of the others have $n = 6$. Also *C. stevioides* and *C. fremontii* are endemic to deserts of relatively recent origin, while *C. glabriuscula* occurs in older mesic habitats.

Aneuploid reduction in chromosome number also occurred in *Haplopappus* (Asteraceae). *H. gracilis* has the lowest chromosome number in flowering plants, $n = 2$. This species clearly is derived on this basis and because the base number for most species in the genus is $n = 4$. Jackson (1962) discovered the putative progenitor of *H. gracilis* in the form of *H. ravenii*, which previously had been unknown. Interspecific hybrids showed that at pachytene the four chromosomes of *H. ravenii* pair with the two of *H. gracilis* almost throughout their entire lengths. At diakinesis various multivalent configurations were observed, and at metaphase I the most frequent association was two bivalents and two univalents. Jackson concluded that the latter evolved from *H. ravenii* or a very similar taxon by a process of unequal translocations and loss of centromeres.

Another possible example of an aneuploid decrease in chromosome number is in *Coreopsis*. *C. nuecensis* has a haploid number of $n = 6$ or 7, and *C. nuecensoides* has a haploid number of $n = 9$ or 10 (Smith, 1974). The former is thought to be derived because it has a subset of the allozymes present in its more variable congener and is less heterozygous (Crawford and Smith, 1982). Five alleles are missing. Hybrids between *C. nuecensis* and *C. nuecensoides* have complex multivalent formations suggestive of multiple translocations (Smith, 1974).

If the lowest chromosome number in a genus is derived, then we can point to other examples of a descent in number. Perhaps the best involves Togby's (1943) study of *Crepis*, where *C. fuliginosa* ($n = 3$) is thought to be derived from *C. neglecta* ($n = 4$). The chromosomes in each complement have distinctive sizes and landmarks. Pairing configurations show that the species differ by one translocation between the first (largest) and fourth (smallest) chromosomes of *C. neglecta* and by another translocation between the second and third chromosomes. The species also differ by three inversions.

There are relatively few examples of chromosomal rearrangements leading to an increase in chromosome number. A well-known one involves *Clarkia biloba* ($n = 8$) and *C. lingulata* ($n = 9$; Lewis and Roberts, 1956). The latter ostensibly is derived from *C. biloba* on the basis of its chromosome number that is further removed from the base number of the genus ($n = 7$) than is the number of *C. biloba*. This putative phylogeny is supported by the observation that *C. lingulata* has a subset of the allozyme alleles of *C. biloba* (Gottlieb, 1974). Moreover, *C. lingulata* is known from only two sites only the southern periphery of *C. biloba*'s range.

Two separate translocations have been involved in the differentiation of *C. lingulata*. One involves an exchange between two chromosomes that results in the formation of a ring of four chromosomes in F_1 hybrids. The other translocation produces a chain of five chromosomes in hybrids. This multivalent contains two pairs of homologous chromosomes and an additional chromosome that has one arm of one pair and one arm of the other. The extra chromosome thus produces two chromosome arms in a triplicate condition in *C. lingulata*. The two species also differ by an inversion whose location is not known.

ANEUPLOID SERIES

Many genera have a series of chromosome numbers generated by the addition or subtraction of single chromosomes. Some examples of such aneuploid series are listed in table 3.1. They are interpreted by their describers as ascending or descending. These series may be quite long, as in the genera *Lapeirousia* and *Haplopappus*. In certain genera (e.g., *Clarkia* and *Phacelia*), it appears that some phylads are ascending whereas others are descending.

One of the more bizarre aneuploid series occurs in *Claytonia virginica*. The following diploid numbers have been reported by Rothwell (1959) and Lewis (1970): 12, 14, 17–20, 22, 24, 26, 28, 30–32, 34 36, 41, 48, 72. This series arose from a combination of aneuploidy and polyploidy. The original base number of the species may have been $x = 8$ (Rothwell, 1959). Aneuploidy even is found within populations such as one at Carthage, Texas, where chromosome number varies from $2n = 24$ to $2n = 41$. Moreover, the modal number varies from year to year (Lewis, 1976).

In most genera, the putative direction of chromosome number change is based on indirect evidence rather than on a molecular phylogeny. Fortunately, there are some series that can be interpreted in light of a molecular phylogeny. Consider first ascending aneuploidy.

The haploid numbers in the genus *Hypochaeris* range from three through six (Cerbah et al., 1999). ITS sequence variation for nuclear rDNA was used to construct a phylogenetic tree that showed that the basal number is four and that chromosomal change was largely one of addition (figure 3.13). It is notable that sister groups may differ by more than one chromosome. On the one hand, *H. cretensis* and *H. oligocephala* have three chromosomes. On the other hand, *H. illyrica, H. maculata,* and *H. uniflora* have six chromosomes. Five chromosomes evolved independently in two separate phylads.

Table 3.1.   Examples of Genera with Aneuploid Series

Genus	Haploid numbers	Reference
	PROGRESSION DESCENDING	
Crepis	6,5,4,3	Babcock (1947)
Phacelia	11,10,9,8,7	Cave and Constance (1947, 1950), Constance (1963)
Clarkia	7,6,5	Lewis (1953)
Lesquerella	8,7,6,5	Rollins and Shaw (1973)
Calycadenia	7,6,5	Baldwin (1993)
Lapeirousia	10,9,8,7,6,5,4,3	Goldblatt and Takei (1993)
Moraea	10,9,8,7,6,5	Goldblatt and Takei (1997)
Hypochaeris	6,5,4	Cerbah et al. (1999)
Bunium	11,10,9,8,7,6	Vasiléva et al. (1985)
Lobelia	14,12,11,10,9,8,7,6	Stace and James (1996)
Sideritis	22,21,20,18,17	Barber et al. (2000)
Haplopappus	9,6,5,4,3,2	Raven et al. (1960), Jackson (1962)
	PROGRESSION ASCENDING	
Hemizonia	9,11,12,13	Kyhos et al. (1990)
Crocus	12,13,14,15	Brighton (1978)
Clarkia	7,8.9	Lewis (1953)
Phacelia	11,12,13	Cave and Constance (1947, 1950), Constance (1963)
Allium	7,8,9	Levan (1932, 1935)
Sideritis	18,19,20,21,22	Barber et al. (2000)

Ascending aneuploidy also appears to have been the primary theme in *Clarkia*. Based on a tree inferred from the nucleotide sequences of the *PgiC* gene, the basal number is $n = 7$ (Gottlieb and Ford, 1996). The highest number reached is $n = 9$.

Descending aneuploidy occurs in the genus *Calycadenia*, which has $n = 4-7$. These numbers are mapped onto the molecular phylogeny of Baldwin (1993) based on the ITS sequence variation for nuclear rDNA. This phylogeny is congruent with one based on the external transcribed spacer of 18S–26S rDNA (Baldwin and Markos, 1998). These phylogenies suggest that $n = 7$ is the base number for the genus. The species with $n = 4$ (*C. spicata*) shares a common ancestor with a species with $n = 7$ (*C. villosa*). Species with $n = 5$ and 6 are in a different phylad.

Hybrids between the more basal species (*C. mollis, C. truncata,* and *C. villosa*) display primarily univalents at meiosis, while hybrids between the more derived

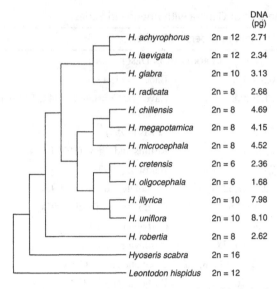

		DNA (pg)
H. achyrophorus	2n = 12	2.71
H. laevigata	2n = 12	2.34
H. glabra	2n = 10	3.13
H. radicata	2n = 8	2.68
H. chillensis	2n = 8	4.69
H. megapotamica	2n = 8	4.15
H. microcephala	2n = 8	4.52
H. cretensis	2n = 6	2.36
H. oligocephala	2n = 6	1.68
H. illyrica	2n = 10	7.98
H. uniflora	2n = 10	8.10
H. robertia	2n = 8	2.62
Hyoseris scabra	2n = 16	
Leontodon hispidus	2n = 12	

Figure 3.13.   ITS consensus tree of *Hypochaeris* species with their chromosome numbers and 2*C* DNA contents. Redrawn from Cerbah et al. (1999), with permission of Blackwell Science Ltd.

species (*C. hooveri, C. spicata, C. multiglandulosa, C. hispida, C. oppositifolia, C. ciliata, C. pauciflora,* and *C. fremontii*) have a high incidence of multivalents and bivalents (Carr, 1975, 1977). Thus, it appears that species with more recent ancestry have the most similar chromosomal complements.

Recurrent descending aneuploidy has been demonstrated in Australian *Podolepis,* as indicated by DNA sequences of the nuclear ITS region and of the chloroplast *matK* gene (Konishi et al., 2000). The descending series begin with $n = 12$ and go to $n = 10$ and $n = 9$ in one lineage, to $n = 8$ and $n = 7$ in another lineage, and $n = 11$ and $n = 3$ in a third lineage (figure 3.14). Chromosomes of five *Podolepis* species are portrayed in figure 3.15.

Recurrent descending aneuploidy also occurs in *Brachyscome.* From a molecular phylogeny based on the sequence of the chloroplast gene *matK,* Watanabe et al. (1999) found that the ancestral base number was $n = 9$. The direction of numerical change is consistent across lineages. Chromosome number declined from $n = 9$ to $n = 8$ and then to $n = 4$ in one clade, to $n = 7$ and then to $n = 6$ twice, and then to $n = 5$ once in another clade, to $n = 4$ and then $n = 3$ in a third clade, to $n = 5$ and then $n = 3$ in a fourth clade, and to $n = 2$ in a fifth clade (figure 3.16). The decrease in chromosome number was accompanied by an increase in karyotype asymmetry. Changes from perennial to annual habit occurred in five lineages. In four of them, this change was accompanied by a reduction in chromosome number.

Both ascending and descending aneuploidy have may have occurred in many lineages. One lineage, *Sideritis* (Lamiaceae) subgenus *Marrubiastrum* is espe-

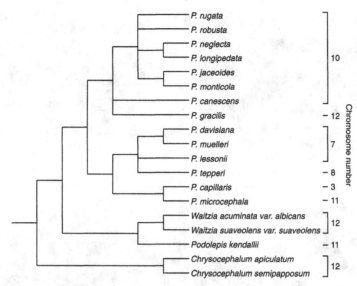

Figure 3.14. Diploid chromosome numbers in Australian *Podolepis* mapped onto a phylogenetic tree based on ITS and *matK* sequences. Redrawn from Konishi et al. (2000), with the permission of CSIRO Publishing.

cially interesting because it is endemic to the Macaronesian Islands and because chromosome numbers have been rather static on oceanic islands (Stuessy and Crawford, 1998). The aneuploid series in this group is the most extensive in oceanic island plants. The genus has nine different haploid numbers. Barber et al. (2000) constructed a phylogeny based on chloroplast DNA (cpDNA) restriction site differences onto which the chromosome numbers are mapped. The genus falls into two major phylads. In one the basal number appears to be 44, with that number dropping to 40, 36, and 34. In the second phylad, the basal number ostensibly is 36, with that number increasing to 38, 40, 42, and 44. Clearly species sharing the same chromosome number may not be closely related.

CHROMOSOMAL DIFFERENCES AS MEASURES OF RELATIONSHIPS

Before the advent of molecular markers, there was a temptation to use chromosome numbers and chromosome pairing relationships in F_1 hybrids as indicators of the evolutionary affinities of the parental species (Seberg and Petersen, 1998). Presumably the more similar the genetic constitution of species the more similar their chromosomal complements would be. Thus, diploid species with different chromosome numbers might be considered to be more divergent than species with the same numbers. Species whose chromosomes did not pair normally in F_1 hybrids often were considered to be more remote than species whose hybrids had normal pairing.

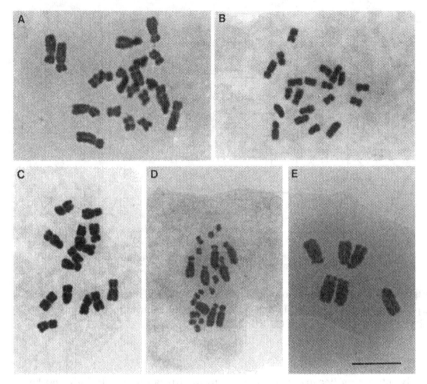

Figure 3.15. Chromosomes of *Podolepis jaceoides* (2*n* = 20), *P. canescens* (2*n* = 20), *P. davisiana* (2*n* = 14), *P. microcephala* (2*n* = 22) and *P. capillaris* (2*n* = 6). Scale bar = 10mµ. Courtesy of Kuniaki Watanabe.

Before including cytogenetics in phylogenetic formulations, consider why the chromosomes of diploid species do not form bivalents in hybrids. First there is the matter of chromosome number. Differences in chromosome numbers arise as a result of reciprocal translocations, centric fusions, and centric fissions. Thus, differences in number need not signify distant relationships. Two diploid plants may differ in chromosome number and yet have exactly the same genotype.

The fallacy of assuming that species with different chromosome numbers are less closely related than those with the same number is illustrated in the molecular phylogeny of *Hypochaeris* (figure 3.13). We see that some species with disparate chromosome numbers (e.g., 6 and 10) are more closely related than are some species with the same number. We also see that species with similar genome sizes may be only remotely related.

Reciprocal translocations and inversions may cause meiotic abnormalities and reduced fertility in the hybrids of species with the same chromosome numbers. However, the level of abnormality in and of itself does not signify the

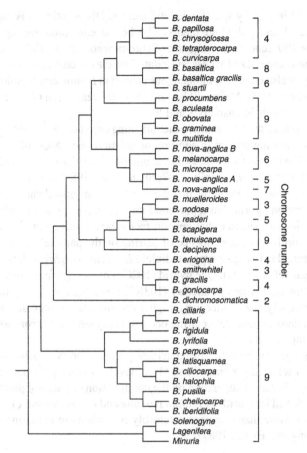

Figure 3.16. A phylogenetic tree of *Brachyscome* species upon which chromosome numbers have been superimposed. Some species have been omitted for the sake of brevity. Redrawn from Watanabe et al. (1999), with permission of the Botanical Society of Japan.

level of genetic divergence between species, because chromosomal rearrangement is not synonymous with genetic differentiation.

For the sake of argument, let us assume that two species have the same chromosomal architecture. Does this mean that chromosome pairing, then, will be a function of the overall similarity of their genomes? It does not, because less than 1% of the DNA appears to be involved in the mutual recognition of homologous chromosomes (Wettsein et al., 1984; Loidl, 1994). This recognition seems to be controlled by association sites whose number and distribution along chromosomes is unknown.

The use of chromosome pairing data in phylogenetic studies was questioned several decades ago by Darlington (1932, 1937) on the basis that pairing may be determined in part by specific genetic factors. His skepticism is supported in both diploid and polyploid plants. The best-documented pairing control genes are the *Ph1* gene on the long arm of chromosome 5B and *Ph2* on the short arm of chromosome 3D of wheat and related species (Benavente et al., 1998). Genetic factors also control homoloeogous pairing in *Lolium* amphiploids (Jenkins and Jimenez, 1995). More will be said about the regulation of chromosome pairing in polyploids in chapters 8 and 9.

In diploid plants, the presence of genes affecting chromosome pairing has been deduced from pairing studies in addition or deletion lines. Most of the studies involve hybridization between diploids and polyploids (e.g., *Agropyron,* Charpentier et al., 1988; *Triticum,* Dvorák, 1987; Shang et al., 1989; *Hordeum,* Thomas and Pickering, 1985). They show that the identity of the added chromosome(s) in later generation hybrids has a profound impact on pairing relationships.

Another line of evidence that genetic factors are involved in chromosome pairing in hybrids is the dependence of pairing on the parental genomes. This is well illustrated in two genotypes of diploid *Hordeum vulgare* × *H. bulbosum* hybrids (Zhang et al., 1999). One hybrid (102C2) has a high frequency of bivalents at metaphase I, whereas another (103K5) has a high frequency of univalents. Correlatively, the former has a higher recombination frequency than the latter. GISH demonstrates that pairing occurred only between chromosomes of different complements.

If there is a genetic basis to pairing in interspecific hybrids, then hybrids from parental lines selected for "high" and "low" levels of pairing should vary predictably in pairing level depending on the combination of parental plants. This, indeed, was found in hybrids of *Lolium perenne* and *L. temulentum* (Taylor and Evans, 1977). More than one gene ostensibly controls meiotic pairing in these hybrids (Aung and Evans, 1987).

OVERVIEW

Chromosomal evolution via rearrangements, especially translocations, has accompanied morphological evolution in a multitude of genera. This was evident many years ago in the divergent karyotypes of species and from chromosome pairing irregularities in interspecific hybrids. However, the nature of the rearrangements often was unclear. Only through the application of comparative genome mapping have we begun to truly appreciate the number of chromosomal alterations by which species differ and the magnitude of these differences.

Chromosome pairing analyses in interspecific hybrids has revealed that many species differ by multiple translocations. In general, as the number of chromosomal differences increases the fertility of hybrids declines. The nature of chromosome pairing relationships in species hybrids has been advanced through the study of synaptonemal complexes. Not surprisingly, chromosome pairing is more compelling in hybrids between close relatives than between distant relatives.

Translocations have led to aneuploid chromosome number series in many genera. The direction of chromosome number change was a matter of conjecture before mapping chromosome numbers onto molecular phylogenies. Now we can see, at least in a few genera, the probable direction of change. Most of the examples are of descending aneuploidy.

There has been a temptation to use chromosome numbers and chromosome pairing relationships in interspecific hybrids as indicators of species affinities. However, as our understanding of the genetic control of chromosome pairing has increased and as we map chromosome numbers onto phylogenies, it is becoming increasingly evident that species affinities are not well portrayed in chromosomal data. Nevertheless, pairing relationships do provide insights into the potential for gene exchange between species.

4

Chromosomal Barriers to Interspecific Gene Exchange

Chromosomal differences between species and the attendant meiotic abnormalities they impose on interspecific hybrids reduce the potential for interspecific gene exchange between species at the same ploidal level. This reduction is the product of partial sterility, on the one hand, and reduced recombination in rearranged chromosomal segments, on the other. The magnitude of the barrier to gene exchange depends on the nature, number, and size of the rearrangements.

HYBRID STERILITY

In general, the fertility of hybrids declines as the number of chromosomal differences between species increases. Large rearrangements have a greater impact on fertility than do small ones. Hybrid fertility is expected to be a function of the percentage of chromosomes that are paired. The more univalents and multivalents that occur the lower the fertility, with a few rare exceptions.

The relationship between chromosome pairing and fertility is illustrated in hybrids between *Gilia modocensis* and *G. malior* (Grant, 1966a). As shown in figure 4.1, fertility is below 20% unless more than 70% of the chromosomes are paired. The pattern of increasing fertility with increasing bivalent formation is broadly representative of what one finds in species hybrids.

I now turn to the impact of translocations on fertility. This may be seen in its simplest form by comparing the fertilities of F_1 hybrids of species that differ by one translocation with the fertilities of F_1 hybrids of species without this rearrangement. Such a comparison was made by Schank and Knowles (1964) on hybrids of *Carthamus* species (Asteraceae). All species have 10 pairs of chromosomes. Twelve types of hybrids were obtained with one translocation, and nine types with none. Pollen fertility in the hybrids with one translocation averaged 40%. Pollen fertility in the hybrids without a translocation averaged 80%. Achene fertility in the translocation hybrids averaged 44% versus 77% in the nontranslocation hybrids.

What is the effect of an increasing number of translocations on hybrid fertility? This question is well addressed in the Hawaiian silverswords. Carr and Kyhos (1981, 1986) estimated the number of translocations in F_1 hybrids from 48 species combinations. The number of translocations varied from zero to three. The mean pollen fertility in hybrids with three translocations was 10% versus 34% in hybrids with two translocations, 72% in hybrids with one translocation, and 96% in hybrids with no translocations (figure 4.2). The relationship between translocation number and fertility is quite striking ($r = -0.49$, $p = 0.01$).

The relationship between translocation number and pollen stainability is not always compelling, as seen in *Helianthus*. Chandler, Jan, and Beard (1986) studied 34 kinds of F_1 hybrids. Some were heterozygous for as many as six translocations and for as few as none. Fertility varied from a few percent to over 50%. Hybrids heterozygous for three of more translocations had lower fertility on average than those with none or few. The correlation between the number of translocations and pollen fertility was $r = 0.30$, which falls short of statistical significance ($p = 0.07$). Pollen fertility in *Helianthus* also is reduced by inversion heterozygosity and the occasional failure of chromosomes to pair.

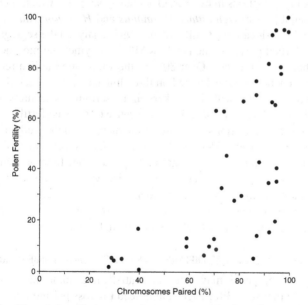

Figure 4.1. The relationship between chromosome pairing in pollen mother cells and pollen fertility in hybrids between *Gilia malior* and *G. modocensis*. Redrawn from Grant (1966a), with permission of the Genetics Society of America.

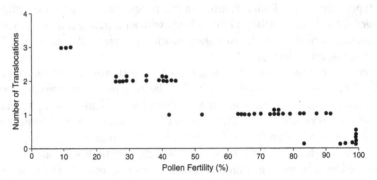

Figure 4.2.   The relationship between the number of translocations by which species differ and the fertility of their F_1 hybrids in the Hawaiian silversword alliance. Based on Carr and Kyhos (1986).

The number of translocations by which species differ may have a rather small impact on hybrid pollen fertility, if chromosome disjunction from multivalents tends to be alternate. This is the case in *Gaura* sect. *Gaura* (Carr et al. 1986a). Hybrids heterozygous for one or two translocations had about 63% fertility versus about 52% in hybrids heterozygous for three or four translocations and 39% for five. Alternate disjunction was a significant mitigating factor.

We know very little the specific contributions of chromosomal rearrangements to fertility reductions in hybrids. One notable study for which such information is available is in *Helianthus. H. annuus* and *H. argophyllus* differ by two reciprocal translocations. Quillet et al. (1995) analyzed the segregation of randomly amplified polymorphic DNA (RAPD), isozyme, and morphological markers in backcross progeny. Over 80% of the variation in pollen fertility is explained by genetic intervals located on three linkage groups. Meiotic anomalies are tightly correlated with the markers circumscribing these intervals, indicating that the translocations were the prime cause of reduced fertility.

One of the first applications of chromosome mapping in the study of hybrid sterility involved *Lens ervoides* and *L. culinaris*. The species differ by a reciprocal translocation. Using a segregating F_2 population, Tadmor, Zamir, and Ladizinsky (1987) were able to correlate four isozyme markers with quadrivalent formation during prophase I and identify the translocation end points. Plants with pollen viability below 65% were heterozygous for the translocation, whereas those with viability above 85% were homozygous.

THE EFFECT OF GENOME SIZE DIFFERENCES ON CHROMOSOME PAIRING

As discussed above, species may differ substantially in genome size. It is tempting to conclude that such differences might lead to gross pairing irregularities in interspecific hybrids and, as such, prevent gene exchange between these species. As I discuss below, the chromosomes of species with divergent genome sizes may or may not pair normally in hybrids. When the latter occurs, one does not know whether it is due to genome size alone or to other factors.

The genus *Zea* offers prime examples of substantial genome size disparities between species that are, nevertheless, accompanied by complete bivalent formation in their F_1 hybrids. The hybrids in question involve *Z. luxurians* (2C DNA content = 9.0 pg) and *Z. diploperennis* (20.5 pg), on the one hand, and *Z. luxurians* and *Z. mays* ssp. *parviglumis* (5.9 pg), on the other (Poggio et al., 1998). Normal chromosome pairing also occurs in hybrids between *Tradescantia crassifolia* and *T. tepoxtlana*, despite the fact that the former has about 1.7 times more DNA per nucleus than the latter (Martinez, 1976). Bivalents alone also have been reported in hybrids between *Allium cepa* and *A. fistulosa*, even though the genome of the former is 27% larger (Rees and Jones, 1967). Pairing occurred even though loops and overlaps occurred between 10% and 60% of the chromosome length.

Normal pairing does not necessarily occur in close relatives with disparate genomes sizes. Seal and Rees (1982) studied the effect of genome size difference on chromosome pairing in hybrids between *Festuca drymeja* and *F. scariosa*, which differ in DNA content by 52%. Each chromosome of the former had about 0.17 pg more DNA than each chromosome of the latter. Pairing was effective in only three of seven bivalents in F_1 hybrids. The smaller homoeologues were more prone to univalent formation than larger ones.

A subsequent study on *Festuca* involving hybrids among seven taxa provides some specific insights into the relationship between genome size difference and chromosome pairing. Morgan et al. (1986) crossed species whose DNA content varied from 3.4 pg/nucleus to 7.17 pg/nucleus. They found a very strong nega-

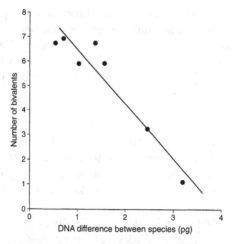

Figure 4.3.   The relationship between the difference in DNA content between species and the mean number of bivalents in their F_1 in *Festuca* sect. *Montanae*. Redrawn from Morgan et al. (1986), with permission of the National Research Council of Canada.

tive correlation between DNA differences between species and the number of bivalents formed at pachytene (figure 4.3). The hybrids of species with a modest difference in genome size had nearly perfect (7 II) chromosome pairing, whereas the hybrid involving a parental difference of more than 3 pg averaged only one bivalent per cell.

Attempts have been made to gain greater understanding of pairing relationships in hybrids by analyzing the synaptonemal complex (SC). The SC is primarily a proteinaceous structure that is formed by the side-to-side union of chromosomal axial elements through transversal filaments at zygotene (Loidl, 1991). Jenkins (1985) analyzed the SC in hybrids between *Lolium temulentum* and *L. perenne*. Each chromosome in the former has about 0.15 pg more DNA than each chromosome of the latter. There was complete and continuous pairing from telomere to telomere in 40% of the bivalents without loops or overlaps. In the other bivalents, chromosome pairing was complete from telomere to telomere but incorporated lateral loops. "Perfect" pairing is achieved by adjusting chromosome length differences both before and during synapsis. Pairing at pachytene was "perfect" most often in larger chromosomes.

As might be expected, chromosome pairing in hybrids between genera with disparate genome sizes is more irregular in hybrids between congeneric species. Consider, for example, hybrids between *Festuca drymeja* and *Lolium multiflorum*. These species differ in DNA content by 66%, the former having the larger genome. Synaptonemal complex formation was quite irregular, with multivalents, foldback pairing, and pairing between nonhomologues (Thomas and Morgan, 1990). None of the chromosomes exhibited complete telomere to telomere pairing or even long lengths of pairing. Bivalent formation was rare.

INTRODUCTION AND THE RESTRICTION OF RECOMBINATION

Structural differences between homologous chromosomes typically lead to reduced pairing and/or genetically unbalanced gametes, which in turn reduce effective recombination rates within rearranged linkages (Hanson, 1959a; Tadmor et al., 1987). This will inhibit or disrupt introgression within and adjacent to the rearranged chromosomal segments (Hanson, 1959b; Jackson, 1985; Sybenga, 1992).

Recombination rates

Interspecific gene exchange can be enhanced by increasing recombination rates between parental linkage blocks (Hanson, 1959a,b; Stephens, 1961; Wall, 1970). One or more generations of selfing or sib-crossing of the F_1 progeny prior to backcrossing should be more effective than backcrossing alone in moving alleles across linkage groups.

Rieseberg et al. (1996) determined how the number of generations of sib-mating before backcrossing affects the level of introgression from *H. petiolaris* into *H. annuus*. They employed three mating designs, as follows: design I, P-F_1-BC_1-BC_2-F_2-F_3; design II, P-F_1-F_2-BC_1-BC_2-F_3; design III, P-F_1-F_2-F_3-BC_1-BC_2. They found that with design I 45% of 58 *H. petiolaris* markers in the collinear portion of the genome introgressed in at least 1 of 56 progeny versus 60% with design II and 62% with design III. Turning to rearranged portions of the genome, only 3.6% of the markers introgressed with design I versus 8.6% with design II and 15% with design III. Thus, their results were consistent with theory.

Reduced rates of recombination associated with chromosomal introgression have been demonstrated in cultivated plants. For example, Rick (1969) introgressed five chromosomes (numbers 2, 6, 8, 10, 11) of *Solanum pennellii* into *Lycopersicon esculentum* in a program involving one chromosome per stock. Reduced recombination occurred in the lines with introgressed chromosomes 8, 10, and 11 relative to *L. esculentum* controls. For chromosome 8, the reduction was 75%. Both species have 12 pairs of chromosomes and meiosis in hybrids is normal. Thus, reduced recombination cannot be attributed to chromosomal rearrangements, unless they are cryptic.

Liharska et al. (1996) studied the effects of alien segments of chromosome 6 on recombination frequencies in tomato. Seven wild lines from *L. pennellii* and *L. peruvianum* served as sources of the introgressed segments. There was a negative correlation between the size of the introgressed *L. peruvianum* segment and the recombination rate of markers flanking the segment. Even an *L. peruvianum* segment as small as 750 kb was sufficient to cause a twofold reduction in recombination frequencies in the *tl-yv* interval, which is much larger. The suppressive effect of introgression was greater for segments from the more remotely related *L. peruvianum* than it was for comparable segments from *L. pennellii*. Similarly, an introgressed segment on tomato chromosome 1 from the distantly related *L. hirsutum* suppressed recombinant frequencies more than a similar-sized *L. pimpinellifolium* segment (Balint-Kurti et al., 1994).

Recombination rates were assessed from genetic maps in backcross progeny of *Lycopersicon esculentum* × *Solanum lycopersicoides*. Chetelat et al. (2000) found that recombination in the BC_1 plants was reduced by about 27% relative to that in the parental species. Recombination suppression was observed for all chromosomes except 9 and 12 in both the distal and proximal regions even though genetic maps of the parental species were essentially collinear. The most severe reduction (70%) was on chromosome 10, and recombination was eliminated on its long arm presumably due to an undetected rearrangement(s). The suppression of recombination between the parental genomes is due primarily to overall genomic divergence rather than chromosomal rearrangements, as suggested in the frequent formation of univalents in F_1 hybrids and about a 20% reduction in chiasma frequency (Rick, 1951; Menzel, 1962).

An assessment of first generation backcross progeny revealed that maize × teosinte hybrids have reduced recombination for markers on the long arm of chromosome 1 (Williams et al., 1995). The level of reduction varied with type of teosinte involved in the cross. An inversion apparently is responsible for the recombination "shrinkage."

Reduced recombination rates also have been associated with chromosomal introgression. One example is in *Gossypium hirsutum*, where a chromosome from *G. raimondii* was substituted for one of its own (Rhyne, 1958). Reduced recombination also follows the introduction of an alien chromosome segment in barley (Görg et al., 1993).

If recombination rates are reduced in F_1 hybrids of chromosomally divergent species, then the map distances between specific markers should be less in hybrids than in the parental species. The extent to which this is the case was determined in a hybrid between the dihaploid *Solanum tuberosum* ($n = 12$) and diploid *S. spegazzinii* ($n = 12$). Kreike and Stiekema (1997) measured map distances between between 14 combinations of markers distributed across 9 of the 12 chromosomes. Ten of the 14 segments were shorter in the hybrids (by an average of 43%) than in *S. tuberosum*. The linkage maps published by Bonierbale et al. (1988) and Tanksley et al. (1992) from hybrids also are considerably shorter than the Gebhardt et al. (1991) linkage map of pure *S. tuberosum*.

Chromosomal introgression

Given that alien chromosome segments can be backcrossed into a species, we may ask about the length of these segments and whether the distribution of introgressed segments is random along the chromosomes. Some insights into these issues are obtained from backcross-inbred lines derived from *Helianthus annuus* and *H. petiolaris*. Rieseberg et al. (1996) found that the mating design affected the length of introgressed fragments in rearranged linkages as well as their transmission frequency. The more generations of sib-mating before backcrossing the larger the length of the introgressed fragments from *H. petiolaris* into *H. annuus*. With one generation of sib-mating, the mean fragment length was 30 centimorgans (cM). With two generations it was 80 cM, and with three generations of sib-mating it was 104 cM.

Additional insights into the length of introgressed segments were obtained from backcross-inbred lines derived from *Lycopersicon esculentum* × *Solanum lycopersicoides*, where the former was the recurrent parent (Chetelat and Meglic, 2000). Introgressed segments were in either homozygous or heterozygous states. The transmission and fixation of *S. lycopersicoides* segments varied for individual chromosomes. There was, for example, a greater number of introgressed segments from chromosome 12, the shortest of the complement, than for chromosome 1, the longest. Most recombinant chromosome segments were short (figure 4.4). The modal class was 1–10 cM, and the majority of segments were less than 30 cM. A few segments exceeded 80 cM, which constitutes a larger portion of an entire chromosome. The largest introgression was chromosome 6, that is, 100 cM long. Somewhat unexpectedly, a majority of the *S. lycopersicoides* introgressions (78%) were in terminal positions. Ninety-one percent of the homozygous introgressions were terminal. Almost all of the introgressed segments less than 15 cM long were terminal.

The question then arises as to the fate of introgressed segments. Do they remain the same size, or do they become smaller over successive generations through recombination and selection? Eshed et al. (1992) provide an answer for *Lycopersicon esculentum* lines containing small chromosome segments from

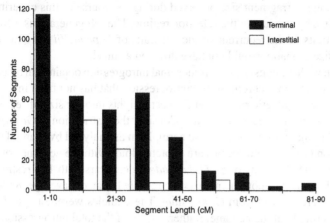

Figure 4.4. Frequency distribution of recombinant chromosome segments according to genetic length in backcross-inbred *Lycopersicon esculentum* × *Solanum lycopersicoides* derivatives. Redrawn from Chetelat and Meglic (2000), with permission of Springer-Verlag.

L. pennelli. Their program entailed backcrossing interspecific hybrids to the former, and then selfing populations for six generations accompanied by strong selection for cultivated tomato phenotypes. The distributions of the introgressed segments for BC_1 and BC_1S6 generations from 120 introgression lines are depicted in figure 4.5. There was a dramatic decline in the length of introgressed segments during the selfing generations.

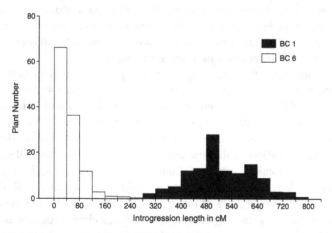

Figure 4.5. Frequency distribution of the length of *Lycopersicon pennellii* introgressions (cM) in the BC_1 generation and the BC_1S_6 introgression lines derived from an *L. esculentum* × *L. pennellii* cross. Redrawn from Eshed et al. (1992), with permission of Springer-Verlag.

The change in fragment size observed during the course of this experiment was, in part, a by-product of the selection regime. The self generations were selected for traits of the recurrent species. If traits of *L. pennellii* were favored, fragment size probably would not have dropped as much.

Thus far, we have considered chromosomal introgression obtained in breeding programs. What about chromosomal introgression that has not been imposed? Plants with dense genetic maps offer the best insights into the size of segments transferred from one species to another. Consider the distribution of *Gossypium hirsutum* chromatin in *G. barbadense* germplasm as analyzed by Wang, Dong, and Paterson (1995). Both species are allotetraploids with the genome composition of AADD. A survey of 54 *G. barbadense* cultivars with 106 restriction fragment length polymorphism markers revealed that on average 8.9% of the alleles were derived from *G. hirsutum*. These alleles were not distributed randomly throughout the genome. Rather, 57.5% of the total introgression was accounted for by five chromosomal regions that span 9% of the genome. The highly introgressed regions were equally distributed across the A and D subgenomes. In the A genome, one region on chromosome 1 had a mean length of 38.9 cM; another on chromosome 5 had a mean length of 30.8 cM. In the D genome, one region on chromosome 14 had a mean length of 19.8 cM, whereas another on chromosome 25 had a mean length of 38 cM. The fifth segment, whose subgenomic location is unknown, had a mean length of 42.2 cM.

Given that chromosomal introgression may occur between species, we would like to know about the relationship between genome disparity and the nature and level of chromosomal introgression. We may expect that the more disparate the genomes the fewer and smaller these segments are likely to be, because a greater proportion of genes from the nonrecurrent species will be ill fit in an alien genetic background and culled by selection. Unfortunately, there is no database with which to test this expectation. Using the same argument, we may expect the level of introgression to be less in rearranged portions of related genomes than in collinear portions of these genomes.

Evidence supporting the latter supposition was presented by Rieseberg, Linder, and Seiler (1995). The subject was *Helianthus annuus* into which chromosome segments of *H. petiolaris* were introduced. The production of generated F_1 hybrids was followed by two generations of backcrossing into *H. annuus* and then by two generations of sib mating. Forty percent of *H. petiolaris* markers in the seven collinear portions of the genome introgressed in at least one of the progeny, with a total genomic coverage of about 40%. Conversely, only 3.6% of the *H. petiolaris* markers from the ten rearranged linkages were transferred. These markers covered less than 2.4% of the structurally disparate chromosomal regions. The results of this study agree with the theory that chromosomal structural differences lower rates of introgression within those portions of the genome.

STABILIZED PRODUCTS OF CHROMOSOMAL REARRANGEMENT

There are two avenues for obtaining chromosomally unique entities via hybridization independent of polyploidy. The first is introgression. When alien chromosome segments are favored, they become an integral part of the recipient's chromosome complement. The second avenue for the genesis of unique entities involves the union of two partially recombined chromosome complements. This is the prime mechanism of recombinational speciation (Grant, 1981). An intermediate entity is formed that carries chromosomal elements of both parental species in greater balance than that obtained with introgression.

The most informative data set on the chromosomal architecture of a stabilized hybrid derivative is for *Helianthus anomalus,* which is derived from *H. annuus* and *H. petiolaris.* As discussed above, Rieseberg, VanFossen, and Derochers (1995) generated linkage maps for *H. annuus* and *H. petiolaris.* They found that the chromosome complement of *H. anomalus* contains components of the

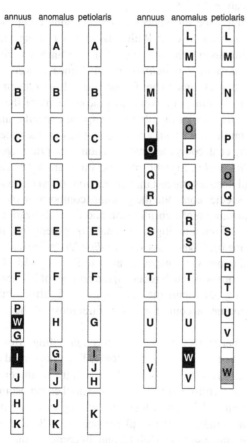

Figure 4.6. Inferred structural relationships of the chromosomes of *Helianthus annuus, H. petiolaris,* and their hybrid derivative *H. anomalus.* The black areas within chromosomes indicate inversions specific to *H. annuus,* and the gray areas inversions specific to *H. petiolaris.* Translocated chromosomes have more than one letter designation. Redrawn from Rieseberg et al. (1995), with permission of Macmillan Magazines, Ltd.

parental genomes, with those of *H. annuus* favored. However, the chromosome complement of *H. anomalus* does not simply combine the complements of its parental species. Rather, it has undergone a minimum of three chromosomal breakages, three fusions and one duplication (figure 4.6). First generation hybrids with *H. annuus* have pollen fertilities of 2–4%, and those with *H. petiolaris* have pollen fertilities of 2–58%.

To study the processes accompanying the origin of *H. anomalus,* Rieseberg et al. (1996) synthesized three hybrid lineages using a combination of advanced generation hybridization and backcrossing. *Helianthus annuus* was the recurrent species. The lineages are as follows: lineage I, $P-F_1-BC_1-BC_2-F_2-F_3$; lineage II, $P-F_1-F_2-BC_1-BC_2-F_3$; lineage III, $F_1-F_2-F_3-BC_1-BC_2-BC3$. Plants from the final generation of each hybrid lineage were surveyed for 197 RAPD markers of known genomic location.

Whereas there might be differences in the genomic compositions of the three lineages, this was not the case. Rather, there was a strong concordance in genomic composition among the lineages. If in one lineage a given chromosome had a linkage block of *H. petiolaris,* the other lineages were likely to have the same block on that chromosome. Chromosomal rearrangements limited the frequency of *H. petiolaris* markers in some portions of the genome. Of particular interest is the similar distribution of species-specific markers in the synthetic lineages on the one hand, and *H. anomalus* on the other ($r = 0.68$; figure 4.7). Rieseberg et al. concluded that selection is the prime agent governing the character of *H. anomalus*. Significant associations among unlinked markers in each of the hybrid lineages suggest that interactions between coadapted genes within the parental species constrain the genomic composition of *H. anomalus*.

Another classic example of hybridization and the stabilization of chromosomal novelty comes from a multigenerational experiment on *Gilia* (Polemoniaceae; Grant, 1966a,b). Grant crossed *G. malior* ($2n = 36$) and *G. modocensis* ($2n = 36$) and obtained vigorous but nearly sterile F_1 hybrids. Chromosome pairing in the F_1 hybrids was much reduced, the number of bivalents per pollen mother cell averaging 6.0 versus the maximum of 18. Structural differences between the two genomes account for some, if not most, of the pairing irregularity.

The F_2 and F_3 generations were obtained through selfing. Most of the latter were weak. One of the healthier and more fertile F_3 plants served as the parent of a lineage (Branch III) that was followed through the tenth generation. As a result of selection for fertility, the frequency of paired chromosomes increased from the F_2 generation to the F_6, by which time pairing was normal (figure 4.8).

The chromosome number of the F_1 plants was $2n = 36$ and of the F_2 was 37–40. The plant giving rise to Branch III had 39 chromosomes. However, its F_4 descendant, which became the progenitor of all later members of Branch III, had $2n = 38$. Pollen fertility in Branch III, which ranged from 13% to 69% in the F_4 generation, tended to increase during subsequent generations reaching from 35% to 89% in the F_{10} as the chromosome number remained at 38 (figure 4.9).

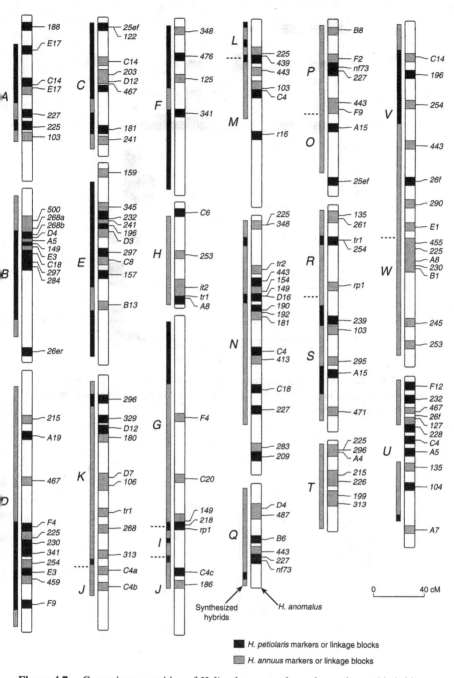

Figure 4.7. Genomic composition of *Helianthus anomalus* and experimental hybrids. The letters to the left of each linkage group designate major linkage blocks in this species and show their relationship to homologous linkages in the parental species. The distribution of parental markers within the *H. anomalus* genome is indicated by the bars within the linkage groups. Areas without information about parental origin are white. Redrawn from Rieseberg et al. (1996), with permission of the American Association for the Advancement of Science.

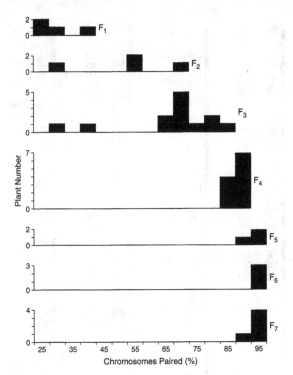

Figure 4.8. The frequency distribution of plants with different percentages of their chromosomes paired in a hybrid *Gilia* line. Redrawn from Grant (1966a), with permission of the Genetics Society of America.

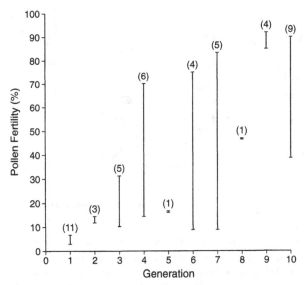

Figure 4.9. The range of pollen fertilities among sister plants in successive generations of selection and inbreeding in a *Gilia* hybrid line. The number of plants assayed is in parentheses. Redrawn from Grant (1966a), with permission of the Genetics Society of America.

Branch III effectively is a "new species" with a new combination of parental traits. It resembles *G. modocensis* in four traits and *G. malior* in three traits and is intermediate in four others. Branch III is reproductively isolated from its progenitors.

OVERVIEW

Chromosomal differences between species are prime factors in prohibiting or retarding gene exchange between congeneric species. They do so by causing or contributing to hybrid sterility and by reducing the level of recombination in the rearranged segments and adjacent regions. The magnitude of the barrier to gene exchange depends on the nature, number, and size of rearrangements. In general, larger numbers of rearrangements are accompanied by greater reductions in fertility. However, if meiotic chromosomal disjunction is alternate, translocation heterozygosity may not impose a reduction in fertility, regardless of the number of translocations.

In addition to being caused by chromosomal rearrangements, pairing irregularities (primarily univalents) in hybrids may be caused by large differences in the size of homoeologous chromosomes. In general, the smaller the difference in interspecific genome size and thus chromosome size, the less pairing is affected. Differences less than 25% may not disrupt the formation of bivalents.

The use of dense genetic maps has made it possible to estimate the magnitude of chromosomal introgression. Given that few maps are available for wild plants, most of our knowledge about chromosomal introgression centers on a few cultivated species, especially on *Lycopersicon*. Here we have indication of introgressed segment size and their location on chromosomes. We also know that segment size declines with successive generations of backcrossing when the traits of the recurrent parent are favored. If the traits of the nonrecurrent parent were favored, at least some of the alien segments would be at a selective advantage, and some of the sizes would be retained. Indeed, genetic maps would allow one to predict which alien chromosome segments would remain intact given selection for certain traits.

The stabilization of hybrid derivatives between species differing in multiple chromosome arrangements has been demonstrated in a few genera. These derivatives may have the same number of chromosomes as their progenitors or a different number.

Dense genomic mapping in *Helianthus annuus* and *H. petiolaris* has made it possible to characterize the origin of chromosome segments in their hybrid derivative, *H. anomalus*. The chromosome complement of this species contains elements of both parental species with those of *H. annuus* favored. Rather than simply combining the complements of its parents, *H. anomalus* has several unique arrangements.

The stabilization of chromosomal novelty also has been demonstrated in *Gilia* in an experiment where vigor and fertility were selected in advance generation interspecific hybrids. A true-breeding fertile derivative was obtained that had a diploid chromosome number of 38 versus 36 that characterized both parental species.

5

Permanent Translocation
Heterozygosity

Translocation heterozygosity occurs independent of hybridization in some species, as described in chapter 2. These translocation heterozygotes typically have reduced fertility and are transient, in the sense that upon self-fertilization they generate chromosomal homozygotes as well as heterozygotes. Even if these translocation heterozygotes have some advantage over their homozygous counterparts, they have no mechanism to transmit this state to their progeny. They can be reconstituted only by crossing between plants with different chromosomal end arrangements.

In some self-fertile species, however, chromosomal heterozygosity is not transient but permanent. Some or all members of these species are translocation heterozygotes (see Grant, 1975). The translocations involve entire chromosome arms. During prophase I of meiosis (diakinesis), the translocated chromosomes form rings, which undergo alternate disjunction at anaphase I. As a consequence, the chromosomes contributed by the paternal parent segregate as a block, as do those contributed by the maternal parent. Because there is no segregation of chromosomal homozygotes, self-fertilization will lead to progeny that carry the same translocation heterozygosity as their parent. Heterozygosity may involve as few as two pairs of chromosomes or all of the chromosome complement. If only some of the chromosomes are involved in translocations, genes on the remaining chromosomes may segregate, and progeny will not be identical to their parents. If all of the chromosomes are involved in translocations, an entire set of chromosomes will act as a supergene, and all of the progeny will be genetically identical to their parents.

Species with permanent translocation heterozygosity may contain more than one type of translocation per population and several types of translocations across populations. The number of chromosomes per ring also may vary among individuals within and among populations. Some individuals may have no rings. As noted by Grant (1981), permanent translocation heterozygosity is a property of individuals, not of species.

Permanent translocation heterozygosity is known in at least 57 species (Holsinger and Ellstrand, 1984). Most species are in the Onagraceae, which as discussed above had an unusually high incidence of polymorphisms for translocations within populations and of interpopulation differences in translocations. The genus *Oenothera* has 50 species that are permanent translocation heterozygotes (Raven and Parnell, 1970; Dietrich, 1978; Raven, 1979; Raven et al., 1979). Most of these species form rings involving the entire complement of 14 chromosomes during meiosis; a small number have rings of 8, 10, or 12 chromosomes. *Gaura* has two species (Raven and Gregory, 1972), and *Calylophus* (Towner, 1977) and *Gayophytum* (Lewis and Szweykowski, 1964), one each; they also form rings involving the entire complement of 14 chromosomes. The other permanent heterozygotes are *Gibasis pulchella* (Holsinger and Ellstrand, 1984) and *Rhoeo spathacea* (Sax, 1931) of the Commelinaceae, and *Isotoma petraea* of the Lobeliaceae (James, 1965). In the latter, various intermediate stages occur from no rings to a one encompassing all 14 chromosomes.

Permanent chromosomal heterozygosity can be accomplished by three different mechanisms. These are best understood in *Oenothera*, as reviewed by Cleland (1972) and below. Two of the mechanisms are predicated on a system of balanced lethals. One involves zygotic lethals as in *Oe. glazioviana* (=*Oe. lamarkiana*) in which all 14 chromosomes have reciprocal translocations. One recessive lethal is associated with one set of seven chromosomes that are transmitted through the pollen, and another recessive lethal is associated with the second set of seven chromosomes that are transmitted through the egg. The only viable progeny obtained from selfing are those with one of each type of chromosome set, that is, translocation heterozygotes. Half of the progeny of selfing are chromosomal heterozygotes, and half are chromosomal homozygotes.

Another mechanism for maintaining permanent heterozygosity is based on gametophytic lethals as in *Oe. biennis,* where all 14 chromosomes have translocations. Here, one chromosome complex (alpha) is functional in pollen development, and the other complex (beta) is not, thus resulting in 50% pollen abortion. Conversely, the beta complex is functional during the production of eggs, and the alpha is not. Accordingly, all of the progeny from selfing will be chromosomal heterozygotes. Steiner (1956, 1960) speculated that the gametophytic lethal systems is related to self-incompatibility factors that were relics from an outbreeding ancestor.

The third mechanism for maintaining heterozygosity is selective fertilization. Selective fertilization occurs in some South American oenotheras (Schwemmle, 1968). In this system, an allele in the alpha chromosome complex causes pollen to fertilize only eggs carrying the complementary allele that is combined with the beta chromosome complex. Correlatively, pollen with the beta complex fertilizes only eggs with the alpha complex.

THE GENETIC CORRELATES OF PERMANENT HETEROZYGOSITY

The retention of heterozygosity

Permanent translocation heterozygosity provides a mechanism for the retention and accumulation of genic heterozygosity. Thus, we would expect ring formers to have greater genic heterozygosity than their outcrossing, bivalent forming progenitors or relatives. The magnitude of this difference depends on the level of genetic divergence between chromosome complexes in the ring formers. Products of interspecific hybridization should be more heterozygous than products of intraspecific hybridization.

This rationale may be tested in *Oenothera* subsect. *Euoenothera*. Through an extensive crossing program, Stubbe and Raven (1979) showed that the subsection contained three distinctive genomes, A, B, and C. Bivalent formers have two doses of a given genome, whereas ring formers may have two doses of the same genome (as in *Oe. austromontana* and *Oe. villosa*) or two different genomes (as in *Oe. biennis, Oe. pariviflora,* and *Oe. oakesiana*). Having two doses of the same genome does not preclude chromosomal hybridity. It simply means that the two chromosome complexes are more alike than in plants with different genomes.

Levy and Levin (1975) conducted an analysis of electrophoretic variation at eleven loci in species within *Euoenothera*. They found that the A and B genomes are quite similar and the B and C genomes the most divergent. The mean heterozygosity, proportion of polymorphic loci, and mean number of alleles per locus for seven species are presented in table 5.1, with their genomic formulas. As expected, the highest levels of heterozygosity (~15%) are in the two ring formers with two genomes. Ring formers with one genome have a mean heterozygosity of about 6.5% and bivalent formers a mean of 4.0%.

Table 5.1.   Genetic properties of some *Oenothera* subsect. *Euoenothera* species

Species	Proportion polymorphic loci	Mean number of alleles per locus	Mean heterozygosity	Genetic complexes
Oe. argillicola	0.20	1.40	8.0	CC
Oe. austromontana[a]	0.30	1.40	10.8	BB
Oe. biennis[a]	0.30	1.40	9.5	AB
Oe. elata	0	1.00	0	AA
Oe. oakesiana[a]	0.40	1.50	15.0	AC
Oe. parviflora[a]	0.40	1.55	14.7	BC
Oe. villosa[a]	0.25	1.30	2.8	AA

[a]Permanent chromosomal heterozygote.
Adapted from Holsinger and Ellstrand (1984).

A comparison between a ring former and its likely progenitor can be made in the subsection *Raimannia*. Ellstrand and Levin (1980a) found that the gene frequencies of the ring-forming and self-fertilizing *Oe. laciniata* are very similar to those of the bivalent-forming outcrosser *Oe. grandis*. The proportion polymorphic loci and number of alleles per polymorphic locus are the same (0.3 and 1.6, respectively). The mean heterozygosity in both species is 5.0%. The ring former did not freeze the higher levels of heterozygosity that occur in some members of *Oe. grandis*.

As discussed above, chromosomal heterozygotes derived via interspecific hybridization are expected to differ most from chromosomal homozygotes in genic heterozygosity. This notion is best tested in *Gaura*, where there is good evidence that *G. triangulata* is the product of interspecific hybridization between the self-incompatible *G. suffulta* and the self-compatible *G. brachycarpa*. First, there is morphological evidence; *G. triangulata* is intermediate to its putative progenitors (Raven and Gregory, 1972). Second, this species contains the diagnostic allozymes and flavonoids of both putative parents (Levin and Levy, unpublished observations). The mean heterozygosity across 12 electrophoretic loci is 18% for *G. triangulata* versus 3% for *G. suffulta* and 1% for *G. brachycarpa* (Levin, 1975b). All plants of *G. triangulata* have the same multilocus genotype for the loci studied.

Unlike the unusual case in *Gaura*, the level of genic heterozygosity within a ring former may vary substantially among individuals within the same taxon. We see this in *Oenothera oakesiana*, where the ranges are substantial and includes plants lacking heterozygosity at the loci studied (figure 5.1). One wonders if larger numbers of markers of various types were assayed, whether such large ranges of variation would be observed. If so, do the more heterozygous plants have greater fitness? There have been no studies in permanent heterozygotes relating genic heterozygosity to fitness.

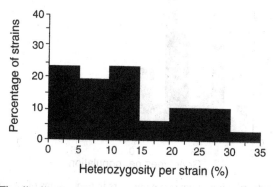

Figure 5.1. The distribution of heterozygosity percentages for strains of *Oenothera oakesiana* collected throughout the species range. Redrawn from Levy and Levin (1975), with permission of the Genetics Society of America.

Population structure

Consider next population structure, which is likely to differ between permanent heterozygotes and bivalent formers. In the latter, all individuals in a population are likely to differ from each other, unless there has been a very long period of intensive inbreeding. This contrasts with paucity of variation expected in populations of permanent heterozygotes. Because they are self-fertile and true breeding, a single individual can found a new population, and should this be the case, all members of that population would have the same genotype.

The genetic diversity within a single local population is dependent on the numbers of founders, the immigration rate, and the age of the population. The more founders, the higher the level of diversity is apt to be, especially if they were from different populations. The high the rate of immigration into an established population the higher the level of diversity is likely to be, especially if the immigrants (seeds or pollen) were from different populations. Because the immigration rate is measured on a per generation basis, older populations are likely to accrue more immigrants, all else being equal, and thus may be more diverse than younger populations.

Chromosomal heterozygotes may cross with others. As each plant produces only one egg and one pollen complex, crosses between two plants will yield identical progeny that differ from both parents. Thus, outcrossing is an additional source of genetic diversity, albeit not a very rich one.

Oenothera biennis populations have very low levels of intrapopulation variation but very high levels of interpopulation differentiation. In a survey of 44 populations using four polymorphic allozyme loci and analyzing 50 plants per population, 28 populations had one genotype, 11 had two genotypes, 1 had

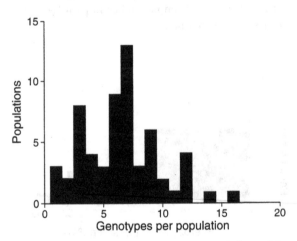

Figure 5.2. The distribution of genotype numbers in populations of *Oenothera laciniata*. Redrawn from Ellstrand and Levin (1982), with permission of the Society for the Study of Evolution.

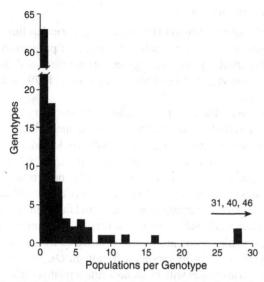

Figure 5.3. The number of *Oenothera laciniata* populations in which given genotypes occur. Redrawn from Ellstrand and Levin (1982), with permission of the Society for the Study of Evolution.

three genotypes, 2 had four genotypes, and 3 populations had the maximum of five genotypes (Levin, 1975c).

   Populations of *Oe. laciniata* also are genetically depauperate but have a somewhat higher level of allozyme variation (Ellstrand and Levin, 1982; figure 5.2). These results agree with the expectations of a model by Holsinger and Feldman (1981), which predicted that any given population should have only one or a few genotypes at equilibrium.

   Both of these ring formers have rather little genetic variation across their populations. A survey of 2,200 plants from 44 populations of *Oe. biennis* uncovered only 47 allozyme genotypes (Levin, 1975c), and a survey of 2,400 plants from 60 populations revealed only 108 such genotypes (Ellstrand and Levin, 1982). Genotypes in both species have restricted distributions. In *Oe. biennis,* a given genotype appeared in an average of 1.5 populations of the 44 studied. In *Oe. laciniata,* a given genotype appeared in an average of 5.2 populations of the 60 studied. Some genotypes did occur in over 30 populations (figure 5.3).

   The aforementioned surveys of *Oe. biennis* and *Oe. laciniata* may have underestimated the amount of variation because of the few markers analyzed. Using allozyme, chromosomal, self-incompatibility allele, and morphological polymorphisms, a minimum of 14 genotypes were detected in one Virginia population of *Oe. biennis* (Steiner and Levin, 1977). It will be most interesting to see what studies using restriction fragment length polymorphisms and microsatellite markers reveal about genetic variation within as well as among populations.

The suppression of recombination

Perhaps the value of chromosomal hybridity in *Oenothera* has little to do with genic heterozygosity per se, but lies in the uniformity of progeny. A particular genotype may do especially well in a given environment and its retention would be favored. This view has been posited by Raven (1979) and Holsinger and Feldman (1981).

The most prominent effect of translocation heterozygosity is the suppression of recombination. Indeed, nearly the entire genome of the complete ring formers in *Oenothera* is sheltered from recombination (Cleland, 1972). This sets the stage for linkage disequilibrium, with their two chromosome complexes carrying different genes. Indeed, the maintenance of gene complexes could be wherein the advantage of chromosomal hybridity lies (Cleland, 1972; Holsinger and Ellstrand, 1984). Because epistasis is an essential factor in the evolution of recombination (Peters and Lively, 2000), so it may be an essential factor in the evolution of gene complexes on chromosomes sheltered from recombination.

Linkage disequilibrium for morphological traits in *Oe. biennis* was established many years ago in crosses with bivalent-forming relatives (Cleland, 1972). When the ring former served as the egg parent, the progeny had a different appearance than when it served as the pollen parent. Electrophoretic studies have also shown that allozyme variants are nonrandomly distributed over the egg and pollen complexes (Levy and Winternheimer, 1977). The magnitude of the difference between these complexes can be expressed as the normalized genetic distance, which is estimated to be 0.04. This is equal to or greater than the distance observed between conspecific plant populations and some intraspecific taxa (Levin, 2000).

In the sexual relatives of chromosomal heterozygotes, selectively neutral alleles are driven toward fixation or extinction because of segregation and recombination. This limits the divergence between allelic sequences caused by recurring mutation. In chromosomal heterozygotes, selectively neutral alleles are not driven to fixation or extinction, because segregation and recombination are absent. Accordingly, we may expect the two genomes of chromosomal heterozygotes to accumulate neutral genetic differences that may be expressed as divergent allelic sequences. Over enough time, the sequence differences between the nonrecombining genomes within a species may become as great or greater than those differences between recently divergent species. The expectation that there may be more sequence divergence between nonrecombining genomes within asexual species than the genomes of related sexual species been realized in the bdelloid rotifers (Welch and Meselson, 2000, 2001).

In addition to neutral divergence between the genomes of permanent heterozygotes, we may expect the accumulation of mutation-generated recessive, deleterious alleles, because their expression is sheltered from selection by normal dominant alleles in the opposite chromosome complex. Harmful nonrecessive alleles also may accumulate if heterozygotes manifest reduced fitness (Rice, 1994). Investigations of asexual genomic haplotypes in animals support these conclusions (Vrijenhoek, 1984). We also may expect the accumulation of pseudo-genes (degraded and nonfunctional genes) owing to the accretion of nucleotide changes, transpositions, and insertion/deletion mutations.

Nonfunctional and harmful genes may drift to "fixation" within a chromosome complex over a very long period of time. Different complexes would have their own unique collection of nonadapted genes. The accumulation of these genes would proceed regardless of whether permanent heterozygosity was gradually accumulated or arose via hybridization between structurally divergent taxa.

In the sexual relatives of chromosomal heterozygotes, selectively neutral alleles are driven toward fixation or extinction because of segregation and recombination. This limits the divergence between allelic sequences caused by recurring mutation. In chromosomal heterozygotes, selectively neutral alleles are not driven to fixation or extinction, because segregation and recombination are absent. Accordingly, we may expect the two genomes of chromosomal heterozygotes to accumulate neutral genetic differences that may be expressed as divergent allelic sequences. Over enough time, the sequence differences between the nonrecombining genomes within a species may become as great or greater than those differences between recently divergent species. The expectation that there may be more sequence divergence between nonrecombining genomes within species than the genomes of related species been realized in the bdelloid rotifers, which include asexual and sexual species (Welch and Meselson, 2000, 2001).

THE EVOLUTION OF PERMANENT TRANSLOCATION HETEROZYGOSITY

The evolution of permanent heterozygosity may proceed along two possible pathways. One involves a gradual accumulation of translocations in outcrossing populations subjected to severe inbreeding (Darlington, 1958; Lewis and John, 1963). One or a few small rings would first be stabilized. One would grow through successive translocations, which eventually could involve the entire chromosome complement.

Recall that some ring-forming members of the *Oe. biennis* complex are homozygous for chromosome complexes. Stubbe (1980) proposed that in these species translocations accumulated gradually. This view is supported by the fact that predominantly bivalent-forming species contain some members with moderate to large rings (e.g., *Oe. grandiflora* and *Oe. maysillesii*). This view also is supported by the fact that some species (e.g., *Oe. wolfii* have two sets of chromosomes that act phenotypically alike (Wasmund and Stubbe, 1986).

Another pathway to permanent heterozygosity involves hybridization between chromosomally divergent races or species. This pathway involves individuals that had a self-incompatibility system. The resulting product would effectively be a stabilized F_1 or advanced generation hybrid. This pathway was embraced by researchers of the *Oe. biennis* complex (Cleland, 1972; Steiner, 1974). Stubbe (1980) also proposed that the taxa with two different chromosome complexes were derived via hybridization between an existing obligate heterozygote with another obligate heterozygote or with a predominantly bivalent-forming taxon. It certainly seems to be applicable to *Gaura triangulata,* which almost certainly is the product of interspecific hybridization.

In South American oenotheras, Dietrich (1978) provides evidence for the gradual accumulation of translocations in some ring formers and for a hybrid origin in others. He argues that species with members spanning the range from no rings to rings involving all 14 chromosomes probably accumulated the large rings by gradual accretion of translocations. Conversely, complex heterozygosity in *Oe. elongata* presumably involved hybridization. One strain even evolved in the last century from a single plant that seemed to be a hybrid between *Oe. longituba* and *Oe. affinis.* This plant is a true breeding complex heterozygote whose progeny does not segregate the parental traits. Other hybrids between these species have rings that do not encompass all 14 chromosomes.

If permanent chromosomal heterozygosity arose gradually, one would hope to find transitional stages between bivalent formers and permanent ring formers. This is indeed what occurs in *Isotoma petraea* (James, 1965). The geographical distribution of chromosome types in southwestern Australia is depicted in figure 5.4. Permanent heterozygosity probably evolved in the cytologically variable Pigeon Rock population that is nested in the area of bivalent formers (James et al., 1990). This population contains a large portion of true breeding ring-of-

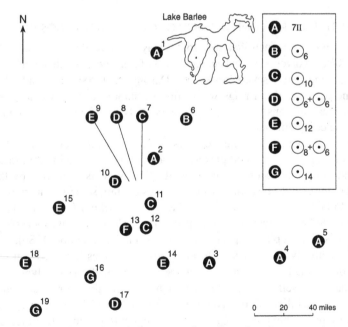

Figure 5.4.   The distribution of chromosomal configurations in western Australian populations of *Isotoma petraea.* Redrawn from James (1965), with permission of Blackwell Science Ltd.

six translocation heterozygotes. These plants combine the standard chromosome end arrangement that is found in all chromosomal homozygotes with one that differs by two translocations. Complex hybridity is thought to have arisen in highly inbreeding lineages in which mutations leading to increasingly efficient transmission of obligately heterozygous parental types were selected. This system presumably then spread to the southwest, perhaps as the species migrated in that direction.

One outcome of permanent translocation heterozygosity in *I. petraea* is the capturing of genic heterozygosity. Based on an analysis of 13 isozyme loci, James et al. (1983) found that the mean heterozygosity for 12 populations without translocations or only floating (impermanent) translocations was 0.02 versus 0.23 in 24 populations of permanent heterozygotes.

The advantage of genic heterozygosity in inbred *I. petraea* is evident in the fact that interpopulation hybrids of chromosomal homozygotes show positive heterosis (Beltran and James, 1974). However, interpopulation hybrids of chromosomal heterozygotes are inferior to their parents, apparently because of poor gene interactions between chromosome complexes.

Regardless of the pathway to permanent translocation heterozygosity, chromosomes must have certain properties or the journey cannot begin. Cleland (1972) suggested that they must be metacentric or nearly so, the centromere must be flanked by heterochromatic (which fractures more readily than euchromatin), and that all chromosomes be near uniform in size. Moreover, chiasmata should be restricted to the distal regions of the chromosomes. Chromosomes with these features occur not only in *Oenothera* but also in the Onagraceae as a whole (Kurabayashi et al., 1962).

With all of the interest in the recurrent origins of auto- and allopolyploids, it is particularly noteworthy that some complex heterozygotes also may have had multiple origins. Dietrich (1978) posits that each of the three subspecies of *Oe. picensis* originated independently from one another, judging from their distinct areas of distribution. Each subspecies has the chromosome complexes of *Oe. odorata* on the one hand, and *Oe. affinis* on the other. *Oe. stricta* also may be polyphyletic. In this case, subspecies have different genomic compositions. Allozyme and cytogenetic evidence suggest that *Oe. laciniata* also had recurrent origins (Hecht, 1950; Ellstrand and Levin, 1980b). *Gaura biennis* may be polyphyletic, because its contains large numbers of different end arrangements present in its putative progenitor, *G. longiflora* (Carr et al., 1986a).

The possibility of the current origins of permanent heterozygotes is intriguing, because until recently it was thought that the recurrent origins of species were restricted to polyploids. Brochmann et al. (2000) provided evidence for the multiple independent origins of the diploid hybrid derivative *Argyranthemum sundingii*. Goodwillie (1999) showed that the self-compatible *Linanthus bicolor* evolved three times from members of different outcrossing clades. The recurrent origins of conspecific ecological races are well documented, as is the recurrent evolution of reproductive system attributes within species and genera (Levin, 2001).

The recurrent or parallel evolution of traits is particular interest in studies of adaptive differentiation. The independent origins of evolutionary novelty within closely related lineages that live in similar environments indicate that evolution has occurred as a result of natural selection imposed by ecological forces (Schluter and Nagel, 1995).

OVERVIEW

In some species translocation heterozygosity is permanent or obligate as opposed to being transient, as it is in so many other species. Most or all members of at least 57 species have this novel condition, and they are self-fertile. The vast majority of these species are members of the Onagraceae. The genus *Oenothera* has 50 permanent heterozygotes, most of which form rings involving the entire complement of 14 chromosomes during prophase I of meiosis. Some have rings of 8, 10, or 12 chromosomes.

Permanent chromosomal heterozygosity is accomplished by one of three mechanisms, all of which occur in *Oenothera.* Two are based on a system of balanced lethals. In one of these systems, one recessive lethal is associated with one set of seven chromosomes that are transmitted through the pollen, and another recessive lethal is associated with the second set of chromosomes that are transmitted through the egg. The only viable progeny obtained from selfing are those that are translocation heterozygotes. In the second system, one chromosome complex is functional only in pollen, whereas the other complex is functional only in the egg. Thus, half the gametes are not viable. All of the progeny of selfing are translocation heterozygotes. The third mechanism for maintaining heterozygosity is selective fertilization. An allele in the sperm causes pollen to fertilize only those eggs that contain the complementary allele associated with the alternative chromosome complex.

Permanent chromosomal heterozygosity provides a means for the retention of genic heterozygosity. Thus, we would expect chromosomal heterozygotes to have higher levels of genic heterozygosity than do related chromosomal homozygotes. This indeed is what is found, although the heterozygosity in the ring forming species never exceeds an average of 15%. One wonders why it is not higher.

Permanent chromosomal heterozygosity also fosters the accumulation of pseudogenes and deleterious recessive genes. This is because all translocated chromosomes, indeed, the entire complement in *Oenothera,* are sheltered from recombination. It remains to be determined the extent to which the incidences of pseudogenes and deleterious recessives are higher in chromosomal heterozygotes than in their homozygous relatives.

Self-fertility allows the founding of a population by a single individual. This coupled with a balanced lethal system means that all members of a population could have the same genotype. Although this extreme has not been described, populations of ring formers have much less genetic variation than do populations of related bivalent formers.

There are two pathways toward the evolution of translocation heterozygosity. One involves the gradual accumulation of translocations in outcrossing populations undergoing severe inbreeding. The other involves hybridization between chromosomally divergent races or species. Both pathways seem to have been taken in *Oenothera*. Indeed, the same pathway may have been taken more than one time per lineage, because some species may be polyphyletic.

6

Polyploidy: Incidence, Types, and Modes of Establishment

Polyploidy, the possession of at least three complete sets of chromosomes, has been a pivotal factor in plant evolution. An increase in ploidal level with or without hybridization often has been associated with speciation and the origin of novel adaptations.

Our perception of the importance of polyploidy in flowering plant evolution has varied considerably over the past half-century. Stebbins (1971) estimated that 30–35% of angiosperms had polyploidy in their ancestry. This was based on chromosome numbers in a genus that are integer multiples of a lower chromosome number. Grant (1963) estimated that 47% of angiosperms were polyploids. He observed that peak haploid numbers were between 7 and 9 and assumed that numbers 10–13 were obtained by ascending aneuploidy. Using estimates of basic haploid numbers in angiosperm families, Goldblatt (1980) concluded that monocots with $n > 10$ (or roughly 70% of all species) were polyploid.

Masterson (1994) compared the guard cell sizes of extinct and extant plants and concluded that about 70% of all species (or those with haploid numbers above 9) were polyploid. The relationship between guard cell size and genome size in the Magnoliaceae is depicted in figure 6.1. In general increasing guard cell size is associated with contemporary polyploidy and is assumed to be a good indicator of ploidal level, time notwithstanding. Thus, if guard cell size were to increase over geological time, it would signify an increase in ploidal level. Masterson showed that such an increase presumably occurred over the past 100 million years in the Platanaceae (figure 6.2).

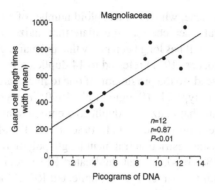

Figure 6.1. The relationship between guard cell size (in micrometers, length × width) and DNA content (pg) in species of the Magnoliaceae. Redrawn from Masterson (1994), with permission of the American Association for the Advancement of Science.

IDENTIFYING POLYPLOIDS

Following the lead of Stebbins, many evolutionists described as diploid the bivalent-forming species of a genus that had the lowest chromosome number that appeared in multiples in other species. The fallacy of this logic was demonstrated in *Helianthus annuus* and *H. laciniatus,* both of which have haploid numbers of 17 and both of which show only bivalent formation. When Jackson and Murray (1983) exposed the microsporocytes of these species to a premeiotic colchicine treatment, they observed meiotic quadrivalents and trivalents. They concluded that some, if not all, of the chromosomes are quadruplicated. The most likely progenitor of *Helianthus* is *Viguiera* ($n = 9$).

There are even species with lower chromosome numbers that are paleopolyploids. One of the best documented is maize ($n = 10$). Maize belongs to the

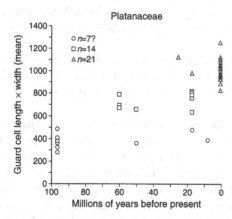

Figure 6.2. Guard cell size (in micrometers, length × width) versus plant occurrence over time in the Platanaceae. Redrawn from Masterson (1994), with permission of the American Association for the Advancement of Science.

grass tribe Andropogoneae, which has a haploid number of 5 (Celarier, 1956). This prompted several researchers to speculate that maize was of tetraploid origin (Anderson, 1945). It has long been know that the maize genome contains duplicated genes. Rhoades (1951) referred to 14 duplicate factors. Premeiotic colchicine treatment leads to the production of one to five quadrivalents, which also is suggestive of polyploidy (Poggio et al., 1990). Isozyme studies were the first to demonstrate that maize has duplicated chromosome segments with collinear gene arrangements (Wendel et al., 1986, 1989). Parallel linkages also have been identified using restriction fragment length polymorphisms (RFLPs). Helentjaris et al. (1988) reported that duplicated RFLP loci occurred on all 10 chromosomes, with the percentage of duplicated RFLP loci ranging from 31% to 60%. A later study of RFLP variation suggested that more than 70% of the loci are duplicated (Ahn and Tanksley, 1993).

The most recent work bearing on maize polyploidy is in the form of DNA sequences of genes from duplicated chromosomal segments. Gaut and Doebley (1997) identified two groups of duplicated sequence pairs, one (group A) more highly divergent than the other (group B). They estimated that sequences in the former diverged about 20.5 million years ago (Mya) versus 11.4 Mya in the latter. The two distinctive groupings of coalescence times indicate that maize is a segmental allopolyploid; that is, it has two homoeologous genomes. The coalescence times for group A sequences estimate the time of divergence of the two ancestral diploid ($n = 5$) species. The coalescence times for group B afford a minimal estimate of the time of interspecific hybridization and chromosome doubling, and an estimate of the onset time of disomic inheritance.

The close ally of maize, *Sorghum bicolor* ($n = 10$), also appears to be an ancient polyploid. Goméz et al. (1998) developed an artificial bacterial chromosome that preferentially hybridized to centromeric regions of 5 of the 10 sorghum chromosomes. This demonstration supports the results of comparative genomic mapping of sorghum and maize in showing that sorghum is a tetraploid (Whitkus et al., 1992). Sorghum diverged from one of the two progenitor lineages of maize about 16.5 Mya, before they united to form polyploid maize (Gaut et al., 2000).

Cotton traditionally has been viewed as an allotetraploid ($2n = 52$). However, a detailed RFLP map revealed that it is a paleo-octaploid, formed between 1 and 2 Mya following the conversion of the base number from 13 to 26 (Reinisch et al., 1994).

The situation is even more complicated in *Brassica*. High-density linkage maps based on RFLP markers show that *B. nigra* ($n = 8$; figure 6.3), *B. oleracea* ($n = 9$), and *B. rapa* ($n = 10$) have genomes containing triplicated copies of an ancestral genome (Lagercrantz and Lydiate, 1996). Thus the species appear to be ancient hexaploids. Collinear regions involving virtually the whole of the three genomes were identified despite the fact that they differ by a large number of rearrangements. The conclusion that the three diploid *Brassica* species are descended from a hexaploid ancestor is supported by the observation that each of these triplicated ancestral genomes is structurally similar to the genome of

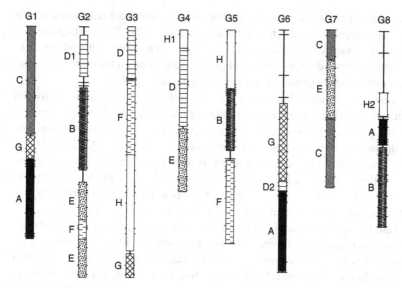

Figure 6.3. Genetic map of *Brassica nigra* showing the distribution of triplicated chromosome segments. Chromosome segments with the same shading share common sets of homologous loci. Redrawn from Lagercrantz and Lydiate (1996), with permission of the Genetics Society of America.

Arabidopsis (Lagercrantz, 1998). Comparative mapping suggests that other crucifers, including *Crambe* and *Sinapis,* also share the common hexaploid ancestor of *Brassica* and thus themselves are hexaploids (Leitch and Bennett, 1997).

The possibility even exists that *Arabidopsis thaliana* is a polyploid, despite its very small genome size and chromosome number ($n = 5$). This is suggested by work of Ku et al. (2000), who compared a sequenced segment of the *ovate*-containing region of tomato chromosome 2 with sequenced portions of the *A. thaliana* genome. The tomato segment was 105-kilobases long. The gene content and order of this segment were similar to those of four different segments of *A. thaliana* chromosomes 2–5. This indicates that these segments were derived from a common ancestral segment through at least two rounds of large-scale genome duplication that may have involved polyploidy. Synteny between loci that map to three linkage groups in soybean also is evidence of duplication in *Arabidopsis* (Gaut et al., 2000).

TYPES OF POLYPLOIDS

Stebbins (1947) recognized three different types of polyploids based on genetic and cytogenetic criteria: autopolyploids, allopolyploids, and segmental allopolyploids. As stated by Stebbins, "autopolyploids usually are characterized by the presence of multivalents at meiosis, of tetrasomic ratios, and in . . . artificially produced [autopolyploids], of slower development and reduced fertility" (p. 423). Conversely, "true allopolyploids may rarely have multivalent associations and tetrasomic ratios, but they usually do not, and they, therefore resemble diploids to a large extent in their cytogenetic behavior" (p. 423). Segmental allopolyploids are somewhat intermediate in that they "will resemble autopolyploids to a greater or lesser degree in possessing multivalents and tetrasomic ratios, but these will be less common [than in autopolyploids]" (p. 423).

In autopolyploids, all genomes are identical or very similar and have the same genomic designation, as in an autotetraploid (AAAA). The corresponding (homologous) chromosomes have equal opportunities to pair during meiosis. In allopolyploids, two distinctive genomes are present (e.g., AABB), and there is no pairing between chromosomes of the different genomes. Thus, multivalents will not be formed. In segmental polyploids, the genomes are partially differentiated (e.g. AAA'A'), so some pairing is possible between homoeologous chromosomes, thus leading to the formation of bivalents or multivalents.

Based on extensive analysis of chromosome pairing behavior in synthetic and natural polyploids, Sybenga (1996) questioned the idea that segmental allopolyploids persist in nature. He contends that although they may be formed by hybridization between related species followed by chromosome doubling, recombination between homoeologous chromosomes eventually will turn them into autopolyploids or less often into allopolyploids.

Instead of basing the classification of polyploids on genetic and cytogenetic data, polyploids can be classified on the basis of their origins. Kihara and Ono (1926) considered autopolyploids to arise within populations, whereas allopolyploids are the products of interspecific hybridization. In essence, autopolyploidy becomes synonymous with intraspecific polyploidy and allopolyploidy with interspecific polyploidy. This approach was adopted by Clausen et al. (1945) and Grant (1981), and by Ramsey and Schemske (1998) in their review of polyploid formation. I prefer to use the genetic–cytogenetic approach, because what constitutes a species is open to debate.

THE PATHWAYS TO POLYPLOIDY

The production of unreduced gametes

Polyploids may arise either through somatic chromosome doubling or through gametic nonreduction (deWet, 1980). The former was considered the prime avenue until Harlan and deWet (1975) showed that the production of diploid (unreduced) gametes was more common than previously believed. Subsequent studies have provided additional evidence that diploid gamete formation ostensibly is the driving force in the formation of polyploids (Bretagnolle and Thompson, 1995; Ramsey and Schemske, 1998).

Both pollen and eggs may be unreduced, but most reports have been of pollen. In part, this reflects the relative ease of recognizing diploid pollen, which usually have diameters 30–40% larger than those of haploid pollen (Ramsey and Schemske, 1998). The differential in grasses may be less (Bretagnolle and Thompson, 1995). The frequency of unreduced eggs typically is estimated from reciprocal crosses between ploidal levels.

In general, there seems to be no significant difference in the frequencies of unreduced eggs and pollen . However, within a given plant, there may be no correlation between $2n$ pollen and $2n$ egg production, perhaps because some mutant genes that influence meiosis have a high sex specificity (Kaul and Murthy, 1985).

The mean production of diploid gametes across flowering plant taxa is 0.56%, excluding hybrids (Ramsey and Schemske, 1998). The frequency of diploid gametes varies widely among species. One of the more extreme cases is *Trifolium pratense,* where all plants sampled had some diploid pollen, the average being about 3% (Parrott and Smith, 1984). In *Solanum commersonii,* 62% of the plants had unreduced pollen (Masuelli et al., 1992), and in *Dactylis glomerata,* 60% of the plants examined had unreduced pollen (Maceira et al., 1992). Conversely, in other species the rate approaches zero.

The percentage of unreduced gametes may vary substantially among conspecific individuals. This is best illustrated in *Trifolium pratense,* in which individual plants ranged in frequency of $2n$ pollen from 1% to 84% (Parrott and Smith, 1984). In *Vaccinium,* individual clones in sect. *Cyanococcus* varied in diploid pollen from less than 1–28% (Ortiz et al., 1992). Variation in diploid gametes is not restricted to pollen. For example, in *Dactylis glomerata,* the percentage of diploid eggs varied from 0.1% to 26% (De Haan et al., 1992), whereas the percentage of diploid pollen varies between 0.1% and 14% (Maceira et al., 1992). In *Solanum,* $2n$ egg production varied from 5% to 23%, whereas $2n$ pollen production varied from 2% to 10% (Watanabe and Peloquin, 1991; Werner and Peloquin, 1991).

Variation within populations for $2n$ gamete formation often is heritable. This is evident in the rapid response of crop strains to selection for $2n$ gamete formation. For example, in three generations of selection on *Trifolium pratense,* the incidence of $2n$ pollen increased from 0.04% to 47%, which amounts to a realized heritability of 0.5 (Parrott and Smith, 1986a). Two cycles of selection on *Medicago sativa* for $2n$ egg and $2n$ pollen production yielded realized heritabilities of 0.60 and 0.39, respectively (Tavoletti et al., 1991).

The genetic basis of $2n$ gamete production is known in several crops as reviewed by Bretagnolle and Thompson (1995). A few examples are mentioned here. Diploid egg production in *Medicago sativa* is under the control of one major recessive gene (*tne1*) and a few minor genes (Calderini and Mariani, 1997). These genes recently have been mapped by Barcaccia et al. (2000). In red clover, diploid pollen formation is controlled by a single gene, whose expression is modulated by a few other genes (Parrott and Smith, 1986b). Three recessive genes operating independently may promote $2n$ pollen formation in *Solanum tuberosum* (Mok and Peloquin, 1975). In *Datura,* a recessive gene (*dy*) controls unreduced egg and pollen formation (Satina and Blakeslee, 1935). Single locus (recessive) control of $2n$ egg formation has been reported in corn (Rhoades and Dempsey, 1966).

One genetic variable that may have a pronounced effect on diploid gamete production is hybridity. By virtue of a high incidence of meiotic irregularities, interspecific hybrids are much more likely to produce diploid gametes than are the species from which they were derived. Ramsey and Schemske (1998) found that on average 27.5% of the gametes of hybrids are $2n$ versus 0.56% in non-hybrids. This is nearly a 50-fold difference!

The environment, as well as genetic factors, may govern the level of unre-duced gamete formation. For example, diploid pollen producers in *Solanum tuberosum* produced between 27% and 35% diploid pollen under warm green-house conditions, whereas during 2 previous years in a cold coastal field they produced 76% and 80% such pollen (McHale, 1983). Some clones of *Solanum phureja* also responded to low temperature with an elevation in unreduced gamete formation (Veilleux and Lauer, 1981). Unusual cold periods increased diploid pollen production in *Datura stramonium* and *Uvularia grandiflora* (Belling, 1925) and in species of *Oenothera* and *Epilobium* (Michaelis, 1928).

The production of diploid pollen grains may be enhanced by stresses other than low temperatures. The frequency of diploid $2n$ pollen in *Achillea mille-folium* was about six times higher in a temperature-cycling growth chamber than in nature (Ramsey and Schemske, 1998). In *Rhoeo discolor,* wide temper-ature variation increased diploid pollen frequency above that afforded by low temperature alone (Sax, 1936).

Low nutrient stress may stimulate $2n$ gamete production as observed in *Gilia,* where diploid F_1 hybrids grown in sand produced about 900 times as many viable tetraploids as hybrids grown in fertilized soil. Parasites (Kostoff and Kendall, 1929) and viruses (Sandfaer, 1973) also may promote $2n$ gamete production. For example, uninfected diploid barley rarely produce spontaneous triploids (0–0.29%), whereas some strains infected by a virus produce nearly 4% (Sand-faer, 1973).

The production of diploid gametes stems from meiotic abnormalities of which there are 13 principal types (Bretagnolle and Thompson, 1995). These abnormalities are related to spindle function, spindle formation, chromosome pairing, and cytokinesis. The best-documented ones involve the failure of the spindle formation during metaphase I or metaphase II. The metaphase I failure leads to first-division restitution of the diploid state, whereas the metaphase II failure leads to second-division restitution. Conspecific individuals may differ in the mechanism of diploid gamete formation, and even within an individual more than one mechanism may function.

The fork in the road

Harlan and deWet (1975) proposed that polyploids may arise through one of two sexual mechanisms. One involves the fusion of two unreduced gametes (bilateral polyploidization). The other involves the fusion of an unreduced gamete with a reduced gamete to produce a triploid (unilateral polyploidization). Through the production of diploid and triploid gametes, triploids can generate tetraploid progeny by either backcrossing to diploids, self-fertilizing, or by crossing with each other as in *Populus tremula* (Johnsson, 1945), *Melampodium* (Warmke and Blakeslee, 1940), and *Dactylis glomerata* (Zohary and Nur, 1959).

Although the genesis of tetraploids via bilateral polyploidization or by the formation of triploid bridges are similarly plausible, the exact pathway responsible for the establishment of any specific naturally polyploid remains a matter of conjecture. There are no specific techniques that we can apply to make an informed choice. Probably some polyploids can trace their origin to one pathway and others to the alternative pathway.

BILATERAL POLYPLOIDIZATION   If bilateral polyploidization is operative in a given situation, then unreduced gametes must be produced in sufficient frequency as to allow a reasonable chance of their union. Such union does indeed happen, although rarely. For example, in open-pollinated diploid apples about four progeny per 1,000 are tetraploid (Einset, 1952). In the diploid *Solanum chacoense,* 2 of 69 plants obtained with self-pollination were tetraploid (Marks, 1966). If one crosses plants that produce diploid gametes more frequently than the norm, the recovery rate of tetraploids is much higher than in open-pollinated seed, as seen in *Costus specious* (Tyagi, 1988) and *Solanum tuberosum* (Jongedijk et al., 1991).

Rather than being dependent on the rare production of $2n$ gametes by the typical plant, bilateral polyploidization may be largely dependent on the presence of a few unusual plants that produce "high" levels of $2n$ gametes (Bretagnolle and Thompson, 1995). For example, in *Anthoxanthum alpinum* most plants have from one to a few percentage of diploid pollen, but a few have more than 20% and a rare plant approaches 40% (Bretagnolle, 2001). Cross-fertilization may provide sufficient opportunity to bring such gametes together, especially if the unusual plants were near each other. The greater the proximity of these plants the greater the likelihood that they would cross (Levin and Kerster, 1974).

Self-fertility would greatly facilitate the union of diploid gametes if diploid eggs and pollen were produced by single plants (Ramsey and Schemske, 1998). In self-incompatible species, diploid pollen may breach the incompatibility barrier, whereas haploid pollen may not. For example, only triploid progeny were recovered after selfing self-incompatible strains of *Pyrus* that produced $2n$ pollen (Lewis, 1949).

UNILATERAL POLYPLOIDIZATION   Unilateral polyploidization requires the production of triploids, and this may be a problem because seeds from the union of n and $2n$ gametes often abort. Abortion usually is due to the failure of endosperm development (Brink and Cooper, 1947). Müntzing (1930a) proposed that the failure resided in the relative ploidal levels of the maternal: endosperm:embryo tissue. These tissues are in a 2:3:2 ratio, respectively, in crosses between diploids, whether they are conspecific or not. In theory, any departure from this ratio would result in seed abortion. When it was shown that many crosses failed to conform to the rule, Lin (1984) proposed an alternative hypothesis that held that normal endosperm development required a 2:1 maternal: paternal genome ratio. Although it worked in many crossing combinations, the

2:1 maternal:paternal genome requirement was violated in some interspecific crosses between ploidal levels.

The most satisfactory concept to explain and predict the outcome of crosses both within and between ploidal levels was developed by Johnston et al. (1980) and is referred to as the Endosperm Balance Number (EBN). The concept holds that normal endosperm development depends on the balance of genetic factors that are contributed by the egg and sperm. Every species has an EBN (or effective ploidal level) that must be in a 2:1 maternal:paternal ratio in the endosperm for crosses to succeed. The EBN of a species is not necessarily equivalent to its chromosomal ploidal level. For example, a diploid species may have a value of 1 or 2. The EBNs are arbitrary values given to species based on their crossing behavior with species with known EBNs. When none are known, a diploid species is assigned a value of two and becomes the standard for a species group. For a cross to be successful, that is, for there to be a 2:1 maternal:paternal ratio in the endosperm, the two entities must have the same EBNs.

If the chromosome number of a taxon is doubled, so is its EBN. Correlatively, an unreduced gamete has an EBN that is twice that of a reduced gamete. This is why $2n$ gametes fail to generate viable triploid progeny when united with n gametes. If the species EBN is two, the endosperm in a seed from a haploid egg and diploid sperm will have a 2:2 ratio. When the egg is diploid and sperm is haploid, the ratio will be 4:1.

The EBN system is under genetic control. Johnston and Hanneman (1996) demonstrated that the control of this system resides beyond one gene and one chromosome in *Solanum* and *Datura*. It is thought that two independent loci may be involved.

The EBN system was formulated with reference to *Solanum* (Johnston and Hanneman, 1980, 1982). Subsequently EBN-like systems have been reported in a few other genera including *Avena* (Katsiotis et al., 1995) and *Trifolium* (Parrott and Smith, 1986c). In *Solanum* and *Avena,* EBN values range from 1 to 4, and in *Trifolium,* from 2 to 8. Species with the same ploidal level may have different EBNs. Values for some species in the aforementioned genera are presented in table 6.1.

A complete triploid block is present in some diploid–tetraploid crossing combinations and not in others, as reviewed by Ramsey and Schemske (1998). They note that when triploids are formed they are a rare occurrence and that the barrier to their formation seems to be stronger when haploid eggs are combined with diploid sperm than when eggs are diploid and the sperm is haploid.

In crosses between species, the stage reached before failure in seed development may be dependent on the ploidal level (and ostensibly EBN value) disparity between species. This is seen in *Sisymbrium,* where species range from diploid to octoploid (Khoshoo and Sharma, 1959). The greater the disparity the earlier in development seeds fail. In reciprocal crosses between ploidal levels, seeds typically reach a later developmental stage when the egg parent has a lower ploidal level.

Table 6.1. Endosperm balance numbers in representative species

Ploidy	EBN	Species
Solanum[a]		
2×	1	*S. chancayense, S. jamesii, S. trifidum*
2×	2	*S. amabile, S. chacoense, S. tuberosum*
4×	2	*S. acaule, S. fendleri, S. papita*
4×	4	*S. curtilobum, S. tuberosum*
6×	4	*S. albicans, S. demissum, S. oplocense*
Avena[b]		
2×	1	*A. clauda, A. eriantha, A. ventricosa*
2×	1	*A. hirtula, A. longiglumis, A. strigosa*
4×	2	*A. barbata, A. marcoccana, A. murphyii*
6×	4	*A. fatua, A. sativa, A. sterilis*
Trifolium[c]		
2×	2	*T. occidentale*
2×	3	*T. diffusum, T. pratense*
2×	4	*T. alpestre, T. nigrescens, T. rubens*
2×	6	*T. pallidum*
4×	4	*T. repens*
4×	6	*T. diffusum, T. pratense*
4×	8	*T. nigrescens*

[a]From Hanneman (1994); Hawkes and Jackson (1992).
[b]From Katsiotis et al. (1995).
[c]From Parrott and Smith (1986c).

RATES OF POLYPLOID FORMATION   Given the possibility of forming tetraploids via two unreduced gametes or via a triploid bridge, we would like to know the relative likelihood of each pathway. Ramsey and Schemske (1998) provide valuable insights based on the incidence of unreduced gametes in diploids and triploids and on triploid fertility in a large number of plants. For the triploid bridge pathway they estimate a rate of autotetraploid formation per generation of 4.98×10^{-5} with selfing and 1.16×10^{-5} with backcrossing. These values compare with 2.16×10^{-5} with selfing or backcrossing for the union of diploid gametes. The rate of allotetraploid formation is estimated to be higher. For the triploid bridge pathway, they estimate a rate of formation per generation of 4.05×10^{-4} with selfing and 3.60×10^{-4} with backcrossing. Formation via the union of diploid gametes is estimated to occur at a rate of 4.05×10^{-2} with selfing and 8.86×10^{-4} with backcrossing. The actual rate of allotetraploid formation will be affected by the incidence of hybridization, the extent to which meiosis is irregular, and the availability of sites suitable for hybrid growth and reproduction.

Whereas we have expected rates for the frequency of spontaneous polyploid formation, almost nothing is known about the actual rate of spontaneous triploidization or tetraploidization in natural populations. A recent study of six diploid populations of *Anthoxanthum alpinum* is particularly noteworthy. Bretagnolle (2001) scored over 6,000 seedlings. This amounts to roughly 1,000 seedlings per population collected from nearly 100 seed parents. Triploid seedlings were recovered in all populations, with incidences ranging from about 0.1% to 0.5%. No tetraploid seedlings were observed. The frequency of unreduced pollen grains was ten times that of triploid plants. These data suggest that the triploid pathway may have been important in the evolution of the tetraploid race of *A. alpinum.*

The presence of polyploids within or near diploid populations does not signify whether the former are products of ongoing genesis or whether they have existed for very long time spans. If genesis were contemporary, polyploids should have no attributes that differ from their progenitors, other than those related to chromosome doubling per se. The *in situ de novo* production of tetraploids in diploid populations seems to be occurring in *Artemisia tridentata* ssp. *vaseyana,* as suggested by the fact that the two cytotypes have identical randomly amplified polymorphic DNA (RAPD) profiles and coumarin content (McArthur et al., 1998).

The *in situ de novo* production of allotetraploids in hybridizing diploid populations has not been observed in nature. Our best understanding of allopolyploid production comes from synthetic hybrids between *Rumex tuberosus* and *R. acetosa* (Swietlinska et al., 1971). First generation hybrids were diploid and vigorous, but with reduced fertility. What is remarkable is that all 20 F_2 plants from the cross *R. tuberosus* $\times$ *R. acetosa* were tetraploid. This situation was made possible by the frequent production of diads instead of tetraploids during meiosis. The gametes from diads presumably were diploid. The pollen grains derived from diads were almost twice as large as those produced by the parent plants.

THE ESTABLISHMENT OF POLYPLOIDS

The production of polyploids occurs within diploid populations. This means that the novel cytotype initially will constitute a very small minority of the total population in which it arises. Occasionally these "founders" replace their diploid counterparts or disperse and establish new monotypic populations. Of concern here is the establishment of polyploid populations in the presence of its diploid counterpart. Specifically, what are the conditions that allow a polyploid subpopulation to grow in its numbers, especially relative to its progenitor, and ultimately replace its progenitor?

Minority disadvantage

Rare tetraploids face a considerable hurdle on the way to locally replacing their diploid counterparts. This is because the vast majority of the pollen they receive will be from diploid parents and because of the difficulty in producing viable triploid seed. If we assume that the two cytotypes have equal numbers of gametes per capita, mate at random and produce abortive triploid seed upon intercrossing, then the incidence of tetraploid × tetraploid crosses and the number of tetraploid seeds per plant will be a function of the proportion of tetraploids in the population (Levin, 1975a). The lower its proportion, the lower its reproductive success relative to its diploid counterpart (figure 6.4), and thus the greater its selective disadvantage. In the extreme, minority cytotype disadvantage would result in tetraploid exclusion. This disadvantage is mitigated to some extent by self-fertilization (figure 6.4).

Sometimes tetraploids overcome a minority disadvantage and spread beyond the site(s) of their origin(s). What conditions would allow this to happen? This issue was first addressed by Fowler and Levin (1984), who studied the competitive relationships of diploid and tetraploid cytotypes by applying the Lotka-Volterra model modified to include parameters accounting for minority disadvantage. They found that the tetraploid may replace the diploid if it had a substantial competitive advantage or as a result of stochastic processes in small populations. If the two cytotypes had divergent ecological requirement, they could coexist. Conditions for these outcomes are quite restrictive. Rodríguez (1996) also used mathematical models to address the roles of competition and niche separation in relation to tetraploid invasibility.

The invasibility of tetraploids in part depends on the frequency with which diploids and triploids produce $2n$ gametes (Felber, 1991, Felber and Bever, 1997). The union of $2n$ gametes generates more tetraploids, which in turn add to the growth of the tetraploid subpopulation much as immigration would. Invasibility also depends on the relative viability and fertility of the tetraploid. Only a substantial advantage will overcome the liability of being rare (Felber, 1991).

In contrast to conjecture about newly emergent tetraploids, it is important to consider what we actually know. Husband (2000) tested the minority disad-

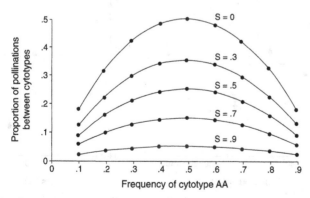

Figure 6.4. Proportions of intercytotype pollination as a function of cytotype frequency with different levels of self-pollination (S). Redrawn from Levin (1975a) with permission of Taxon.

vantage principle in experimental populations of *Chamerion angustifolium* with different proportions of diploids and tetraploids. Tetraploids produce nearly half the flower number produce by diploids but have a small advantage in ovule numbers per flower. The flowering phenologies of the two cytotypes overlapped by 80%. Consistent with minority disadvantage principle, the percentage of seed set per fruit in diploids is a positive function of their frequency (figure 6.5). Oddly enough, seed set in the tetraploids was rather insensitive to its frequency. The basis for this discrepancy remains to be determined.

Assortative mating

Asynchronous flowering times and/or pollinator fidelity and preference also may cause deviations from random mating (viz., positive assortative mating) and thus promote tetraploid invasibility. The recent study by Segraves and Thompson (1999) on *Heuchera grossulariifolia* is particularly informative in these respects. Flowers of tetraploids are significantly larger than those of diploids in a common garden. Sepals of tetraploids ranged in color from light yellow-green to white, whereas in diploids sepals usually were a dark yellow-green to yellow-green. Common-garden tetraploids flowered about 5 days later than diploids and reached flowering peak about 3 days later. Most important, the cytotypes have different suites of pollinators, and 6 of 15 shared pollinators preferentially visited one cytotype (table 6.2).

Asynchrony between ploidal levels has been described in several species. For example, diploid *Plantago media* flowered almost 3 weeks earlier than its tetraploid derivative at a site where the cytotypes were intermixed (Van Dijk

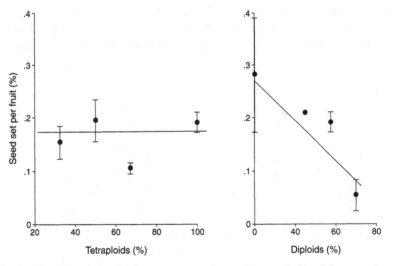

Figure 6.5. The relationship between proportion seed set per fruit and the percentage of tetraploids in populations of *Chamerion angustifolium* for tetraploids and diploids. Redrawn from Husband (2000), with permission of the Royal Society of London.

Table 6.2. Mean pollinator visitation to *Heuchera grossularifolia* per observation period on the Salmon River when both cytotypes were flowering

Pollinator	Diploid mean	Tetraploid mean
Hymenoptera		
Andrena buckelli	0.90	0.80
A. nigrocaerulea	0.60	0.40
Bombus spp.	0.40	0.80
Bombus bifarious nearticus	0.25	1.25[a]
B. centralis queens	0.44	1.38[a]
B. centralis workers	1.20	0.10[a]
Dolichogenidea spp.	1.50	1.50
Lasioglossum spp.	1.68	0.71[a]
Nomada spp.	1.39	0.62
Lepidoptera		
Greya politella	0.25	1.25[a]
Diptera		
Bombyllius major	0.27	1.73[a]

[a]Significant difference between ploidal levels
Source: Adapted from Segraves and Thompson (1999).

et al., 1992). As a result, the overlap in their flowering times was less than 30%. Conversely, tetraploids initiated and completed flowering about 2 weeks earlier than diploids in mixed populations of *Dactylis glomerata* (Maceira et al., 1993). Regardless of which cytotype is first, these flowering differentials impose a significant reproductive barrier between cytotypes and greatly increase the proportion of crosses within cytotypes above that predicted by random mating.

Most models assume that diploids and tetraploids have the same ecological tolerances, but we know that chromosome doubling per se may significantly alter these tolerances (Levin, 1983). Cytotypes may differ in temperature and moisture tolerances and substrate preferences, as discussed in chapter 7. In some instances, tetraploids may outcompete diploids (e.g., *Dactylis glomerata*, Maceira et al., 1993). Given that diploids and polyploids may differ in phenology, pollinator preference, and ecological tolerances and that polyploids may be somewhat self-compatible when their diploid progenitors are not, it is likely that in some species the possibility of polyploid establishment is much greater than we might imagine.

The competitive status of polyploids

Consider next multigeneration experiments involving mixed populations. These experiments are especially informative because they take into account factors operating throughout the life cycle. Hagberg and Ellerström (1959) followed cytotype frequencies within and across generations in variously balanced diploid–tetraploid rye mixtures over a 3-year period. Rye is an outcrossing species. The frequencies of diploids and tetraploids changed little from planting to harvest, regardless of the initial balance. However, the minority cytotype declined in frequency from one generation to another. This occurred because crossing between cytotypes produced abortive triploid seeds, and the proportion of triploid seeds increased as the minority cytotype declined in frequency.

Corn is another outcrossing species in which mixtures of diploids and tetraploids were established at various frequencies. Cavanah and Alexander (1963) found that the frequency of tetraploids declined from one generation to another even when they constituted as much as 90% of the original population. This decline was due to the competitive advantage of pollen from diploids over pollen from tetraploids both on diploid and tetraploid silks and to the differential production of abortive triploid seed that ensued.

The specific competitive advantage of haploid pollen over diploid pollen on tetraploid stigmas is understood in *Trifolium pratense*. Haploid pollen grows at a rate of 41.8 microns per minute in tetraploid styles versus 32.5 μm/minute for diploid pollen (Evans, 1962).

One experiment with ploidal mixtures has been conducted on a selfing species, namely *Arabidopsis thaliana*. Bouharmont and Mace (1972) reported that in a population in which diploids and tetraploids were sown in equal numbers, the proportion of tetraploids increased for five successive generations. The shift occurred because tetraploid seeds have somewhat higher germination rates than diploid seeds and because tetraploids produce more seeds per plant. Since these plants are self-pollinating and self-fertilizing, there was no opportunity for triploid production or for competition between gametophytes of different ploidal levels.

On the one hand, the minority disadvantage models suggest that the local establishment of polyploids may occur only under a very limited set of conditions. On the other hand, in some experiments polyploids did not decline in frequency. Although these experiments were not conducted for extensive periods of time, they do indicate that the models may not have strong predictive value, and perhaps the conditions for polyploid establishment are not as restrictive as they suggest.

Note that each model on polyploid establishment includes some assumptions that may not be valid. For example, most models assume random mating among ploidal levels. This is not necessarily the case for a number of reasons. First, as noted above, microgametophytes of one ploidal level may have a competitive advantage over those of another ploidal level. Second, crosses between ploidal levels may not be compatible. Third, self-incompatibility in tetraploids may not be complete (Stebbins, 1956, 1958). Self-fertilization is the most effective way of propagating tetraploids when they are rare and substantially increases their invasiveness above that obtained with cross-fertilization (Rodríguez, 1996).

Most models assume that diploids and tetraploids have the same ecological tolerances, but we know that chromosome doubling per se may significantly alter these tolerances (Levin, 1983). Cytotypes may differ in temperature and moisture tolerances and substrate preferences, as discussed in chapter 7. In some instances, tetraploids may outcompete diploids (e.g., *Dactylis glomerata*, Maceira et al., 1993). Given that diploids and polyploids may differ in phenology, pollinator preference, and ecological tolerances and that polyploids may be somewhat self-compatible when their diploid progenitors are not, it is likely that in some species the possibility of polyploid establishment is much greater than we might imagine.

THE EXPANSION OF POLYPLOIDS

While attention has been given to the local establishment of polyploids, almost nothing has been written about their expansion. Let us assume that polyploids gradually increase in frequency in a local population in spite of the early minority disadvantage. Does this mean that for them to become a successful population system, they would have to spread to other populations of diploids and in each case overcome minority disadvantage? I think not.

We need not think about the expansion of a polyploid entity only in terms of existing populations. If polyploids and diploids have different ecological tolerances, then polyploids may become established in sites removed from their progenitor, as well as among them. Polyploids never have to replace diploids. They only need to reach numbers sufficient for their propagules to disperse from sites of their origin to others where the diploid is absent. If they can survive and reproduce there, then new populations of the polyploid may be established without opposition. These "pure" populations would be the focal point of dispersal to other sites lacking the diploid. The new entity could thus expand its range quite rapidly.

The spread of polyploid novelty has the same inherent problem as the spread of a novel chromosomal translocation. In the case of the latter, heterozygotes are partially sterile. New translocations suffer a minority disadvantage. The less frequent they are, the lower the probability that novel translocations will be homozygous and the lower the probability that two such homozygotes will interbreed. With partial heterozygote sterility, the only way that the novel translocation will increase is by stochastic processes until it exceeds 50%, unless novel translocation homozygotes have a very substantial advantage over standard homozygotes. Even then it is unlikely that the novel arrangement would diffuse through a population system, because after being introduced into a new population it would still be hostage to stochastic processes (Barton, 1979; Lande, 1985). Rather, its spread ostensibly will be predicated on its association with different ecological properties that allow translocation homozygotes to establish populations removed from those with the standard arrangement (Levin, 2000). It is not surprising that polyploid races are ecologically distinct and often have geographical distributions quite different from their diploid prototypes.

Thus far, the discussion has focused on the establishment of sexual polyploids in diploid populations and their subsequent spread to other sites. Apomixis in polyploids eliminates the disadvantage they face when they are rare. Obligate apomicts are not dependent on cross pollen for reproduction, and thus their reproductive success is uncoupled from their frequency. Facultative apomicts have less of a minority disadvantage than their sexual counterparts. Thus, if apomictic polyploids have a competitive advantage, occupy different niches, or have broader ecological tolerances than their diploid prototypes, they have a much better chance than do sexual tetraploids to establish a subpopulation at the site of their origin and then spread beyond that site.

Apomictic polyploids also may have an establishment advantage because each clone is a distinctive entity that may differ from others in ecological tolerance. If one clone has an advantage over its sexual counterparts or is ecologically divergent, it may spread within a population and to other sites (Vrijenhoek, 1984). Sexual polyploids of different parentages will cross among each other. Thus, they will not avoid the homogenizing effect of recombination, and this may hinder their extension into new niches (García-Ramos and Kirkpatrick, 1997; Kirkpatrick and Barton, 1997).

CONTACT ZONES BETWEEN PLOIDAL RACES

Races of different ploidal levels may make contact along a front a few to several hundred miles wide. Broad contact zones, also referred to as hybrid zones when hybrids are present, are thought to arise from the secondary contact between previously allopatric races (Petit et al., 1999). Narrow contact zones may be the result of secondary contact or the expansion of a newly emergent polyploid population from one or more diploid populations.

Given that ploidal races have different ecological tolerances, contact zones are apt to lie along an environmental transition with one race adapted to conditions at one end and the second race to conditions at the other end. The zone will be maintained by selection against parental types in alien environments and hybrids in parental environments (Barton and Hewitt, 1985; Hewitt, 1988).

The depth of contact zones depends on the spatial pattern of the habitats supporting each race, the fitness liability of populations in alien environments, the extent to which pure populations are interdigitated prior to contact, and the distance that propagules are dispersed (Hewitt, 1993). Depth also will be affected by the mating disadvantage experienced by the minority entity. The greater the liability of being rare and being in an alien environment, the shallower the zone. Similarly, the less the races were interdigitated and the lower the dispersability of pollen and seeds, the shallower will be the zone.

During the past decade, there have been several detailed studies of contact zones with hybrids and without hybrids. An example of the former involves diploid ($2n = 36$) and tetraploid ($4n = 72$) *Chamerion angustifolium* in the Rocky Mountains of southern Montana–northern Wyoming, where diploids occur at the higher elevations (Husband and Schemske, 1998). The spatial heterogeneity of diploids and tetraploids along one section the Beartooth Highway is depicted in figure 6.6. Populations in this area often are composed predominantly of one cytotype. Migration between sites especially suitable for diploids

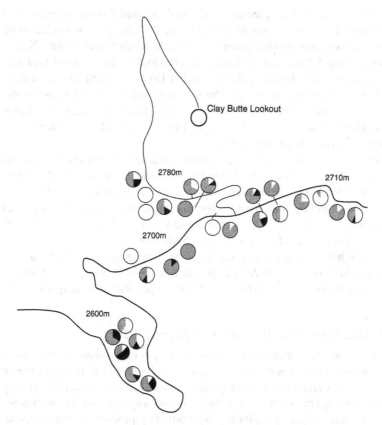

Figure 6.6. The frequency of diploids (white), triploids (black), and tetraploids (gray) in populations of *Chamerion angustifolium* near the Beartooth Highway in the Rocky Mountains of the United States. Redrawn from Husband and Schemske (1998), with permission of the Botanical Society of America.

and sites especially suitable for tetraploids apparently is opposed by selection. Triploids typically are present when diploids and tetraploids are in close proximity. The observed frequency of triploids is 9.2%, which is far below the 48% expected with random mating and equal viability of cytotypes. About 7% of the seeds produced in mixed populations are triploid. The frequency of triploids increases as the ratio of cytotypes approaches 1:1. The hybrid zone is best characterized as a complex mosaic rather than a cline, even though it is maintained by cytotype sorting along an environmental gradient. The hybrid zone is a few kilometers deep.

Another well-studied hybrid zone involves *Galax urceolata*. Diploid races ($2n = 12$) and tetraploid races ($4n = 24$) meet in the Blue Ridge Mountains, from Virginia to Georgia, with diploid populations prevailing toward the north and tetraploids toward the south. Burton and Husband (1999) found that 60%

of the 42 populations sampled had both cytotypes and that the percentage of each varied markedly among neighboring sites. Each cytotype tended to be very rare or very common at a given site, rather than being intermediate. Nearly half of the populations with diploids and tetraploids also contained triploids. Triploid constituted 11% of all plants sampled. On average, triploids were most prevalent when diploids were rare. The hybrid zone is tens of kilometers deep. A relatively deep contact zone (50–100 km) also has been described in European *Lathyrus pratensis* between a large eastern diploid race and a large western tetraploid race (Brunsberg, 1977).

There are two contact zones between diploids and tetraploids, but without hybrids, in *Plantago media:* one in the Pyrenees and the other in the Alps (Van Dijk et al., 1992). Diploid populations occur at lower elevations and have a more southerly distribution than their tetraploid counterparts. Diploid frequencies range from about 45% to 100% within a distance of 300 meters near Formigal. The contact zone in the Pyrenees is about 5 km deep.

Shallow hybrid zones also occur between diploid and tetraploid *Dactylis glomerata* in Europe, North Africa, and the Middle East (Borrill and Lindner, 1971). Near Nazareth (Israel) the zone of mixed populations is only about 100 meters wide.

THE ECOGEOGRAPHY OF DIPLOIDS AND POLYPLOIDS

Polyploids typically replace their diploid prototypes in space and along ecological gradients. This suggests that polyploids usually spread from the periphery of the diploid's niche space. Ecological contrasts between ploidal levels may be quite striking but may vary in character between species. The disparities between diploids and autopolyploids can be direct by-products of chromosomal change and/or by-products of selection on the polyploids. Differences between diploids and allopolyploids may arise as a result of ploidal change or fixed hybridity. Selection also may have contributed to the divergence.

Ecological contrasts between diploids and autopolyploids

Different ploidal levels may be adapted to different temperature regimes and moisture regimes. Divergent tolerances to temperature occur in *Chamerion angustifolium,* where diploids prefer the coldest climates, tetraploids somewhat warmer climates, and hexaploids the warmest climates (Mosquin, 1967). Diploid *C. latifolium* also is more cold tolerant than is its tetraploid counterpart (Small, 1968). Diploids in many other species are more cold-tolerant than tetraploids, as described in *Vaccinium uliginosum* (Hagerup, 1933), *Anthoxanthum odoratum* (Hedberg, 1969), *Fraxinus americana* (Schaefer and Miksche, 1977), *Claytonia cordifolia* (Lewis, 1967), and *Centaurea jacea* (Hardy et al., 2000). This pattern, however, is not universal. Tetraploids in *Hedyotis caerulea* (Lewis and Terrell, 1962), *Suaeda maritima* (Sharma and Dey, 1967), and *Empetrum nigrum* (Hagerup, 1927) are more cold tolerant than are diploids.

An interesting change in soil moisture tolerance occurs in *Eragrostis cambessediana.* Diploids occur in wet habitats, whereas tetraploids occur at the base of nearby dunes and octoploids on very dry dunes (Hagerup, 1932). Conversely, diploid *Galax aphylla* prefers more xeric habitats than does its tetraploid derivative (Baldwin, 1941). In *Tripleurospermum inodorum* (Asteraceae), diploids are adapted to wetter, more maritime conditions than are tetraploids (Kay, 1969), and in *Cruciata taurica,* diploids occur in more mesic habitats than do their derived tetraploids (Ehrendorfer, 1980). In several North American species of *Eupatorium,* polyploids prefer drier, more open habitats than do their diploid prototypes (Watanabe, 1986).

In some species, diploids and polyploids sort out on the basis of whether an area was glaciated or not. The general pattern is for polyploids to occur in glaciated areas, with diploids restricted to unglaciated areas. This dichotomy is seen in *Biscutella laevigata* (Manton, 1937), *Parnassia palustris* (Gadella and Kliphius, 1968), *Saxifraga ferruginea* (Randhawa and Beamish, 1970), *Hedyotis caerulea* (Lewis and Terrell, 1962), and *Tolmiea menziesii* (Soltis, 1984).

Diploids and tetraploids also may prefer different light regimes. In *Deschampsia caespitosa,* diploids are mainly confined to woodlands whereas tetraploids occur in more open habitats (Rothera and Davy, 1986). This pattern also is seen in *Solidago nemoralis* (Brammel and Semple, 1990) and in *Dactylis glomerata* (Lumaret et al., 1987). Conversely, in *Achlys triphylla,* diploids prefer open habitats, and tetraploids shaded ones (Fukuda, 1967).

The alternative patterns of tetraploid divergence may be represented within a single species. For example, in *Dactylis glomerata* the diploids of some subspecies are mainly confined to woodlands, whereas tetraploids prefer open areas (Lumaret et al., 1987). However, in other subspecies diploids occupy more shady habitats than sympatric tetraploids.

Spatial contrasts between diploids and autopolyploids

Given that autotetraploids have ecological tolerances divergent from their diploid progenitors and spread from their sites of establishment, what shall we expect of their geographical distributions relative to each other? We can say with certainty that initially their ranges will overlap, because tetraploids arise in diploid populations. What happens then to a tetraploid's range depends on the pattern of habitat availability. How fast it changes depends on the tetraploids' dispersability and invasiveness. While the tetraploid is expanding, the range of the diploid also may change, because the conditions to which it is exposed are inconstant, as are those of the tetraploid.

The spatial relationships of progenitor and derivative can be reduced to three broad types. The cytotypes may be (1) partially or completely sympatric and ecologically isolated, (2) parapatric, or (3) allopatric. Both sympatry and parapatry offer the cytotypes an opportunity to hybridize.

Sympatry with ecological isolation appears to be the prime spatial relationship of ploidal races (Lewis, 1980). This is well illustrated in *Atriplex confertiflora,* a species endemic to the western United States. The species has 2*n*, 4*n*, 6*n*, 8*n*, and 10*n* chromosome races (Stutz and Sanderson, 1983). The diploid and tetraploid races have the broadest ranges (figure 6.7). The races sort out

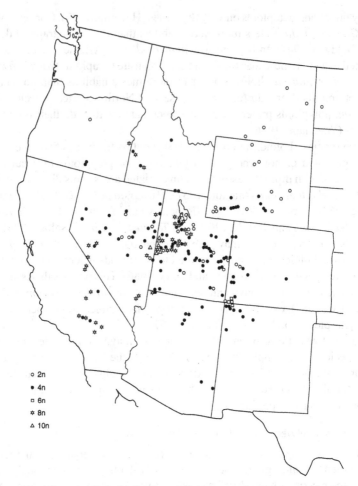

Figure 6.7.   Geographical distribution of ploidal races in *Atriplex confertifolia*. Redrawn from Stutz and Sanderson (1983), with permission of the Botanical Society of America.

in space on the basis of climatic differences. Tetraploid populations grow at lower elevations than do diploids, and higher ploidal populations usually grow at lower elevations than do tetraploids. Stutz and Sanderson (1983) conjecture that polyploid populations in the Great Basin are of recent origin, because they are found in valleys that were filled with water as recently as 12,000 years ago. This is most likely to be true of the octoploids and decaploids, because they typically are restricted to Pleistocene lake-bottom soils.

   Sympatry with ecological isolation also occurs in *Dalea formosa* (Spellenberg, 1981). Tetraploids and hexaploids are largely restricted to the Chihuauhuan Desert, whereas diploids occur there as well as in northern Arizona, New Mexico, and Texas.

Another example of sympatry in the southwestern United States and Mexico involves diploid, tetraploid, and hexaploid *Ambrosia dumosa* (Raven et al., 1968). The hexaploids are confined to the Mojave Desert, whereas diploids and tetraploids extend southward into Baja California and northward into northern and southern Nevada. Several ploidal races of *Atriplex canescens* also are sympatric in the deserts of California and Arizona and adjacent Mexico (Sanderson and Stutz, 1994).

However, not all ploidal races within southwestern American species are sympatric. A case in point is *Larrea divaricata*. Here we see a stepwise westward progression from a diploid race in Texas and New Mexico, to a tetraploid race in Arizona, to a hexaploid race in southern Nevada (figure 6.8).

The parapatric distribution of chromosome races is well exemplified in the European distribution of diploid and tetraploid cytotypes of *Plantago media* (figure 6.9; Van Dijk et al., 1992.). Contact between the cytotypes is achieved in the Pyrenees and in the Alps. The present-day distributions of cytotypes in the Pyrenees may be the result of range expansion from different refugia, probably before 10,000 years before present (Van Dijk et al., 1992).

Although parapatry may be rather local, as in *Plantago media,* it may also be quite extensive, as in *Epilobium angustifolium.* In this circumpolar species, diploids occur at higher latitudes than do tetraploids. In North America, a contact zone extends near the southern limit of the boreal forest and southward into the Rocky Mountains (Mosquin and Small, 1971). *E. latifolium* also has widespread diploid and tetraploid races with borders that approach each other, but they seem not to intermix.

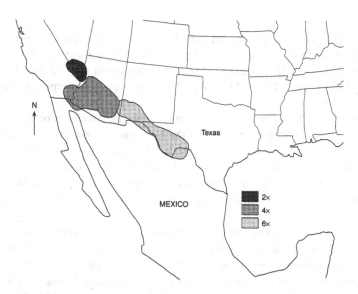

Figure 6.8. Geographical distribution of *Larrea divaricata* ssp. *tridentata* cytotypes. Redrawn from Yang and Lowe (1968a) and Yang and Lowe (1968b), with permission of Madroño.

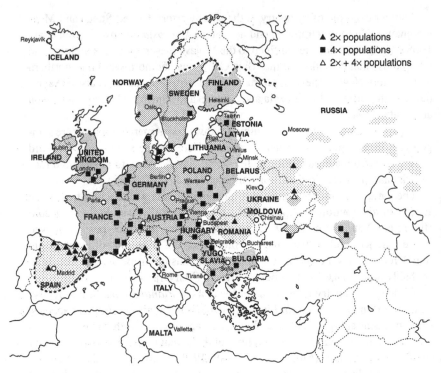

Figure 6.9.   The distribution of *Plantago media* cytotypes in Europe. Redrawn from Van Dijk et al., (1992), with permission of the Linnean Society of London.

Rather few pairs of diploids and polyploids have ranges that are separated by hundreds or thousands miles. Perhaps the most blatant example involves *Nymphoides indica*. This pantropical species is diploid in the Old World and tetraploid in the New World (Ornduff, 1970). The races are morphologically indistinguishable.

When species have more than two ploidal races, the lower numbered races almost invariably are the progenitors of the higher numbered races. Thus, we might expect the higher numbered races to have smaller ranges than the lower numbered ones, at least early in their evolutionary histories. Given enough time, they will reach the geographical limits imposed by their ecological tolerances. Then their ranges may be smaller or larger than those of diploid or tetraploid races.

A survey of the ecogeographical relationships of ploidal races reveals that polyploids often have larger ranges than their diploid progenitors. One example is in *Galium anisophyllum* (Ehrendorfer, 1965). The diploids are limited to well-separated refugia in southern and central Europe that during the Pleistocene were little glaciated or not at all. In each area containing diploids, one finds more widely distributed tetraploids. In Europe, tetraploid *Plantago media* is widely and more or less continuously distributed, whereas the diploids have restricted and disjunct ranges (Van Dijk and Bakx-Schotman, 1997). Also in Europe, the range of *Dactylis glomerata* diploids has been divided and reduced over the past 50 years as a result of deforestation (Lumaret, 1988). Conversely, the ranges of tetraploids have expanded, as they can invade some of the newly created habitats. In the southeastern United States, we find that diploid races of *Eupatorium cuneifolium, E. hyssopifolium, E. leucolepis, E. pilosum,* and *E. rotundifolium* are very restricted, whereas polyploids have broad distributions (Watanabe, 1986).

Ecological contrasts between diploids and allopolyploids

Because allopolyploids are of hybrid origin, we could not rely just on the effects of chromosome doubling to surmise how they might differ from their progenitors. Indeed, because of their hybrid ancestry, we might expect them to be somewhat intermediate between their progenitors in their initial ecological requirements. This expectation would be most fitting if polyploids were derived from F_1 hybrids. Some degree of intermediacy also could be obtained even if polyploids were derived from advanced generation or backcross hybrids.

Ecological intermediacy is not a necessity in allopolyploids. Transgressive segregation in hybrids could lead to substantial niche differentiation in polyploids. Transgressive segregation also could occur at the polyploid level if there were intergenomic recombination. A recent literature survey by Rieseberg et al. (1999) showed that transgressive segregation is ubiquitous in plant hybrids. Examples include transgression for cold, heat, drought, and heavy metal tolerance and for resistance to herbivores and pathogens. Thus, we may conclude that polyploids could become quite divergent from their ancestors due to a combination of ploidal effects and recombination.

Ehrendorfer (1980) and Stebbins (1971) before him observed that polyploids tend to occur in habitats distinctive from both ancestors. This is consistent with the argument that the establishment of polyploids is far more likely if they are ecologically divergent. If there were considerable niche overlap, newly emergent polyploids could be overcome by competition or hybridization, as discussed above.

There are many examples of ecological disparities between diploids and allopolyploids. *Erythronium quinaultense* is an allotetraploid endemic to the Olympic Peninsula of Washington State that grows with neither parent, although both species grow in nearby areas (Allen, 2001). One parent, *E. montanum,* is found in montane and subalpine habitats at elevations from 800 to 2,000 meters and extends near the Pacific Coast northward into British Columbia and southward to central California. The other parent, *E. revolutum,* occurs principally in low-elevation riparian habitats on the east side of the Cascade Mountains in southern Washington and northern Oregon, except for disjunct populations on the Olympic Peninsula and in British Columbia. The polyploid grows in wet, montane, coniferous forests at elevations between 500 and 850 meters.

Ecological divergence between an allopolyploid and its putative parents also has been reported in the orchid *Spiranthes* (Arft and Ranker, 1998). The polyploid *S. diluvialis* occurs primarily along the slopes of the Uintah Mountains (Utah) and Colorado Front Range. One parent, *S. romanzoffiana,* is a montane plant of moist areas from Alaska to Newfoundland and southward into the Rocky Mountains. The other parent, *S. magnicamporum,* is a plains plant that occurs in moist areas through the Great Plains. The parental species are only sympatric in Illinois.

Not all polyploids are distinct from both progenitors. A polyploid's ecological tolerance may overlap that of one parent but not the other. An example of such involves the tetraploid *Gilia transmontana* and its diploid ancestors, *G. clokeyi* and *G. minor* (Day, 1965). *G. minor* occurs in fine sandy outwash plains and in the vicinity of vernal pools in southern California. *G. clokeyi* begins in the eastern Mojave and extends across southern Nevada into Utah and Arizona. This species is found on higher slopes in limestone soils that contain abundant moisture for a considerable part of the growing season (spring). *G. transmontana* occurs in the eastern portion of *G. minor*'s range, the western portion of *G. clokeyi*'s range, and the interval between them. It sometimes forms mixed populations with *G. clokeyi.*

Another example of a polyploid overlapping in ecological preference with one of its progenitors is *Aster ascendens* (Allen and Eccleston, 1998). This species grows primarily throughout the Great Basin in low to middle elevations in sagebrush country, where it prefers wetter than average microsites. One putative parent, *A. occidentalis,* with which it is sympatric and with which it occasionally occurs, tends to grow farther west or at higher elevations in somewhat wetter sites. The other putative parent is *A. falcatus,* which is primarily a Great Plains species that extends from the Northwest Territories southward to the ponderosa pine forests of northern Arizona.

Finally, there are cases where the ecological tolerances of the tetraploid overlaps those of both progenitors. A fine example of this is found in *Tragopogon* (Novak et al., 1991; D. Soltis, personal communication). Diploid *T. dubius* occurs in dry sites, *T. pratensis* in wet meadows, and *T. porrifolius* on semishaded slopes or near ditches at the bottom of hillsides. The polyploids *T. miscellus* and *T. mirus* have broader niche widths than do the diploids. They are most abundant in intermediate habitats and are outcompeting *T. pratensis* and *T. porrifolius* where they grow together. The polyploids occur in western Washington and adjacent Idaho and are spreading rapidly (Soltis et al., 1995).

Another example of a tetraploid's niche overlapping those of both parents involves *Gilia malior* and its presumptive progenitors *G. minor* and *G. aliquanta* (Day, 1965). *G. minor* occurs in fine sandy outwash plains and in the vicinity of vernal pools in southern California, whereas *G. aliquanta* is found on higher slopes in limestone soils that contain abundant moisture for a considerable part of the growing season (spring). *G. malior* is found on granite and basalt formations, but it also is found in habitats of the diploids. *G. malior* is sympatric with both diploids in southern California but also occurs as disjunct populations farther to the north along the California–Nevada border.

An interesting example of a very broad ecological tolerance is seen in the polyploid apomictic *Antennaria rosea,* which has its forebearers among eight closely related sexual diploid species (Bayer, 1997). For the most part, the diploids have distinctive habitat tolerances and tend replace each other as one goes from the steppes to the lower montane, upper montane, subalpine, and alpine zones in the Rocky Mountains (figure 6.10). In contrast to the diploids, *A. rosea* is distributed across the entire spectrum of habitats occupied by the diploids. Bayer proposed that this unusual niche breadth may be the consequence of *A. rosea* having multiple origins, perhaps involving more than one pair of species. He also proposed that hybridization between the polyploid and several diploids could be responsible for ecological diversification in the former.

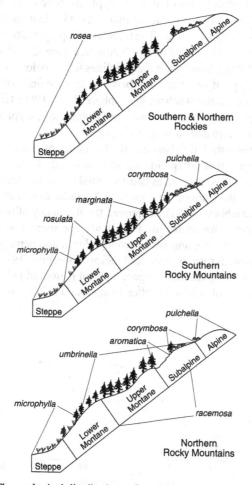

Figure 6.10. The ecological distributions of *Antennaria rosea* and eight of its progenitors in the Rocky Mountains. Redrawn from Bayer (1997), with permission of the Council for Nordic Publications in Botany.

Sometimes the parental species of polyploids do not differ substantially in their ecological tolerances, and the polyploids end up somewhat similar to both parents. As a case in point, consider a species complex in *Clarkia* composed of three diploid and two tetraploid species largely restricted to southern and central California (Lewis and Lewis, 1955; Smith-Huerta, 1986). The diploids are *C. unguiculata, C. epiloboides,* and *C. modesta.* The tetraploid *C. delicata* is derived from the first two species, and the tetraploid *C. similis* is derived from the second two. All of the species are found in oak woodlands.

Transgeneric ecological correlates of polyploidy

Although the genus has been focal point of studies on polyploidy, there also has been interest in discerning transgeneric patterns in geographical and ecological distributions of polyploids, almost all allopolyploids. Studies in the European and arctic flora showed that polyploidy increases with latitude (figure 6.11; Löve and Löve, 1957; Hanelt, 1966). The frequency of polyploids also increases with latitude on the Pacific Coast of North America from California (34%) to British Columbia (56%); however, it then declines to just below 50% in Alaska (Stebbins, 1984). A positive correlation between latitude and polyploidy also appears to be present in the southern hemisphere (Hair, 1966; Hanelt, 1966). Stebbins (1971) concluded that the lowest percentages of polyploidy occur in floras of warm temperate and subtropical regions.

At first blush, the correlation between the incidence of polyploidy and latitude seems to suggest that polyploids are more tolerant of extreme cold than diploids. This is not necessarily so, as demonstrated by the fact that the percentage of polyploids in the flora of the high Alps is similar to that in the foothills and plains below (Stebbins, 1971). Moreover, the frequency of polyploids differs among different growth forms. Polyploids are more common among perennials herbs than among annuals or woody plants, and perennial herbs are the predominant growth form in very cold regions (Stebbins, 1971).

Ehrendorfer (1980) argues that the geographical pattern of polyploid occurrence may be the result of polyploids often being found in labile or successional

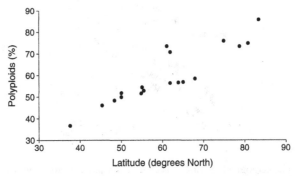

Figure 6.11. The relationship between the percentage of polyploids and latitude in European and arctic floras. Adapted from Löve and Löve (1957) and Hanelt (1966).

biota, whereas diploids often are found in stable habitats or climax assemblages. He showed these tendencies in the floras of lower Austria (table 6.3) and the Ivory Coast.

The types of habitats occupied by polyploid species may have to do with their relative ages. Ehrendorfer (1980) and Stebbins (1985) contend that relatively young polyploids tend to occur in disturbed or seral communities, whereas older polyploids are expected in more stable or climax communities. In addition to the evidence they present, support for this idea comes from the habitat preferences of young and old polyploids in the Sheffield (Britain) flora (Hodgson, 1987). By their definition, younger polyploids are apt to have both parental species extant, whereas older ones have lost one or both parental taxa.

There has been considerable speculation that because polyploids have greater heterozygosity and biochemical diversity than do diploids, they might have broader ecological tolerances than do diploids. Were this the case, polyploids would tend to occur in more geographical regions and in more habitats within regions than diploids.

Stebbins and Dawe (1987) shed light on the first issue in an analysis of the distributions of diploids and polyploids within 75 genera with large European ranges. When they compared the proportions of diploids and polyploids that were widespread in each genus, they found no significant difference between them. Polyploidy has no effect on geographical distribution.

Petit and Thompson (1999) recently compared the habitat breadth of diploid and polyploid species in the flora of the Pyrenees. They classified 451 taxa according to ploidal level and ecological range. Neither the ecological range of genera nor their component species were related to ploidal level. This finding is consistent that of Stebbins and Dawe (1987) in indicating that polyploids do not have broader tolerances than diploids.

The surveys by Stebbins and Dawe (1987) and by Petit and Thompson (1999) involve collections of unrelated species. They do not provide a contrast between related diploid and polyploids. Accordingly, an effect of polyploidy within genera might be confounded, and thus obscured, by the taxonomic variable.

Another relationship long sought regarding polyploidy is that with growth form. Do annuals, herbaceous perennials, and wood plants differ in the occur-

Table 6.3. The pecentages of diploid and polyploid species in plant communities of lower Austria

Community	Stable		Labile	
	Diploid	Polyploid	Diploid	Polyploid
Deciduous forest	72.1	27.9	34.5	65.5
Pine forest	79.1	20.9	49.0	51.0
Rock steppe	68.8	31.2	62.5	37.5
Meadow	81.2	18.8	27.2	72.8

Source: Adapted from Ehrendorfer (1980).

rence of polyploid species? Based on an analysis of 286 genera, Stebbins (1971) found that polyploidy is concentrated in perennial herbs (table 6.4).

Based on the Stebbins survey, we would expect that the rate of chromosome number evolution via polyploidy would be higher in lineages of herbs than in lineages of woody plants. To determine the validity of this expectation, Levin and Wilson (1976) obtained chromosome numbers for 8,378 angiosperm species from the literature. They then estimated the mean increase in chromosome number diversity in typical lineage within a genus.

The mean increase in chromosome number diversity per lineage per million years due to polyploidy was 0.050 in herbs versus 0.010 in shrubs and 0.001 in trees. Thus, the results agree with those presented by Stebbins. The higher level of polyploidy in herbs was paralleled by their higher level of aneuploidy than in the other two growth forms (0.02 in herbs, 0.0005 in shrubs, and 0.0003 in trees). Another noteworthy finding in this study is that the rate of chromosomal evolution via polyploidy and aneuploidy is very strongly correlated (positively) with the rates of species evolution in these growth forms. This is not surprising because chromosome number change is an important driver of speciation.

Polyploidy in relation to genome size

Whereas much attention has been given to the incidence of polyploidy in relation to external environmental variables, internal environmental variables have been ignored. Genome size is one such variable for which an extensive database recently has become available. Is the occurrence of polyploidy independent of genome size? Grif (2000) recently showed that the incidence of polyploidy within families indeed depends their mean genome sizes, with families with larger genomes having less polyploidy. In monocot families with small genomes (7–8 pg/nucleus), the percentage of polyploid species is 86–96%, whereas in families with large genomes (48–79 pg/nucleus), the percentage of polyploids is 20–25% (figure 6.12). Correlatively, in dicot families with small genomes (2–3 pg/nucleus), the percentage of polyploids ranges between 55% and 80%, whereas in families with more than 23 pg/nucleus, the percentage falls to between 5% and 29%. Grif suggested that the negative relationship between polyploidy and genome size reflects an intolerance to very large amounts of DNA.

Table 6.4. The frequency of polyploidy in genera with different growth forms

Growth form	Number of genera	Percentage of genera with polyploids			
		<26%	26–50%	51–75%	>75%
Annual	21	57	24	19	0
Perennial herbs	200	38	31	27	4
Woody	65	65	23	11	1

Source: Adapted from Stebbins (1971).

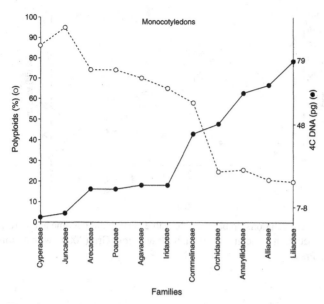

Figure 6.12. The relationship between 4*C* value per diploid genome and the percentage of polyploid species in some monocot families. Redrawn from Grif (2000), with permission of Academic Press.

If small genomes are more conducive to polyploidy than large ones, this pattern should be evident within genera as well as among them. As shown by Grif (2000), grass genera with small genomes tend to have a higher percentage of polyploid species than do those with large genomes (figure 6.13).

THE EVOLUTIONARY SUCCESS OF POLYPLOIDS

The driving forces behind polyploids' ability to thrive in the habitats they now occupy and to expand into habitats beyond the limits of their progenitors are (1) relatively high levels of fixed heterozygosity, (2) novel genotypes, and (3) novel nuclear–cytoplasmic interactions. Most of the emphasis has been placed on the first factor (Thompson and Lumaret, 1992; Soltis and Soltis, 1993; Jiang et al., 1998), but new information suggests that the other two factors also may be quite important.

Fixed heterozygosity

The notion that increased heterozygosity, fixed or transient, provides an advantage to polyploids is supported by experiments on crop plants. First, there was the demonstration in maize that autotetraploids derived from inbred lines and essentially homozygous were much less vigorous than autotetraploids derived from hybrids between such lines that had two alleles per heterozygous locus (Randolph, 1941, 1942). Autopolyploids derived from open-pollinated lines also were much more vigorous than those from inbred lines. Later, Lundquist

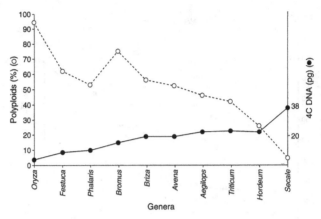

Figure 6.13. The relationship between 4C value per diploid genome and percentage of polyploid species in some grass genera. Redrawn from Grif (2000), with permission of Academic Press.

(1966) showed that in autotetraploid rye, the performances of predominantly tri-allelic and tetra-allelic lines produced by two generations of hybridization involving four different inbred parents were superior to those of di-allelic lines derived from one generation of hybridization. The superiority of the tri- and tetra-allelic condition over the di-allelic condition also was demonstrated in alfalfa autopolyploids (Bingham, 1980).

The only study on wild polyploids relating the level of heterozygosity to plant performance was on the autotetraploid *Dactylis glomerata*. Tomekpe and Lumaret (1991) assessed the association between quantitative traits and heterozygosity at eight electrophoretic loci. They found a positive ($r = 0.15$) and significant (1% level) correlation mean leaf production and multilocus heterozygosity (figure 6.14). Moreover, there was a significant association between the number of alleles and leaf production at two of the loci (*GOT-1* and *PX-1*). Tomekpe and Lumaret (1991) also found a significant positive correlation ($r = 0.13$) between multilocus heterozygosity and panicle number. The number of alleles at *PX-1* was positively related to panicle production.

The possibility of a relationship between heterozygosity and polyploid success was explored in *Draba*. Success was measured in terms of a species' ecological amplitude, broader species being more successful. Brochmann and Elven (1992) assigned each of 16 species an ecological amplitude based on the proportion of its populations present in each of 17 habitats, as expressed by the Shannon-Weiner diversity index. They found that species with higher ploidal levels tended to have broader ecological amplitudes (figure 6.15). They attributed this pattern to the higher levels of heterozygosity associated with plants of higher ploidal levels (figure 6.16).

Allozyme heterozygosity in polyploids may endow plants with greater biochemical diversity than could be present in diploids. In diploids, a dimeric

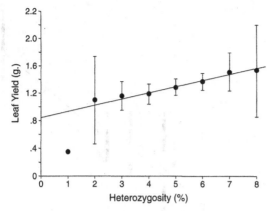

Figure 6.14. The relationship between heterozygosity (based on seven electrophoretic loci) and leaf yield in *Dactylis glomerata*. Redrawn from Tomekpe and Lumaret (1991), with permission of the Society for the Study of Evolution.

enzyme can produce two parental enzymes and one heterodimer. In contrast, a tetraploid with three alleles can produce three parental enzymes and three heterodimers and a tetraploid with four alleles can produce four parental enzymes and six heterodimers. This elevated diversity might increase an organism's biochemical versatility.

Genetic novelty

Genetic novelty is afforded by chromosome doubling per se. The ecological tolerances of organisms may change substantially simply by increasing gene dosage, as discussed in chapter 7. Genetic novelty also may be afforded by transgressive segregation. Transgressive character states, related to complementary

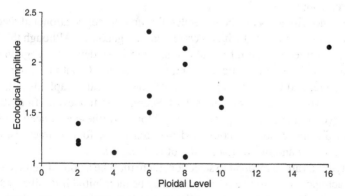

Figure 6.15. The relationship between ploidal level and ecological amplitude in *Draba* species. Redrawn from Brochmann and Elven (1992) with permission of the senior author.

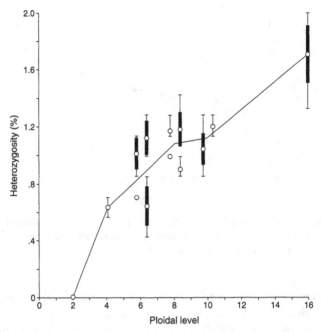

Figure 6.16.   The relationship between ploidal level and heterozygosity at seven electrophoretic loci in *Draba* species. Means are shown as circles; standard deviations, as bars; ranges, as lines. Redrawn from Brochmann and Elven (1992), with permission of the senior author.

genes and to a lesser extent epistasis, have been described in diploid hybrids in many genera (Rieseberg et al., 1999). Some diploid transgressive hybrids could be the substrate for chromosome doubling. Transgressive segregation also could occur after chromosome doubling, if crossing over occurs between homoeologous chromosomes.

Genetic novelty also may be generated through the segregation and recombination of traits associated with divergent diploid genomes. Although this has not been observed in natural populations, there is evidence for segregation and recombination in synthetic allotetraploids in *Gilia*. Grant (1952) crossed *G. millefoliata* and *G. achilleaefolia* and obtained several tetraploids in the F_2 generation, which he subsequently selfed. Segregation for several floral traits was observed in the F_3 and F_4 generations. Moreover, these advanced generation hybrids shared some recombined parental traits, four characteristic of *G. achilleaefolia* and two characteristic of *G. millefoliata.*

If polyploids are to respond to selection, there must be a pool of genetic variation to act upon. One source of variation may be their initial instability, which manifests at the molecular and organismic levels in synthetic allopolyploids. This instability may arise from a manifold change in the genetic environment of divergent genomes wherein newly interacting genes readapt to one another

(Templeton, 1981; Coyne, 1996). The instability of synthetic allopolyploids is discussed in chapter 9 in relation to saltational genome evolution.

Complementary genes in divergent genomes or epistatic interactions of genes in divergent genomes may allow polyploids to respond to selective pressures that their diploid ancestors could not. This may have happened in *Gossypium barbadense* and *G. hirsutum*, both of which have the genomic designation AADD (Jiang et al., 1998). Diploids with the A genome and the aforementioned tetraploids produce seeds with spinnable fibers. Seeds of the D genome diploids do not produce spinnable fibers. Intense artificial selection has consistently produced AD tetraploid cottons whose fibers are much superior to those of A genome diploids. The substantial influence of the D genome on fiber properties of tetraploid cotton indicates that polyploidy may offer unique avenues for responding to selection.

Nuclear–cytoplasmic interactions

Conventional wisdom has it that the success of an allopolyploid is dictated by its nuclear genomic constitution. Very little consideration has been given to the possibility that the chloroplast and mitochondrial genomes also have substantial influence on how polyploids fair. This possibility increases as we begin to appreciate the magnitude of chloroplast and mitochondrial divergence between species. It also increases as we find evidence for strong bias toward certain maternal parents in interspecific hybridizations and allopolyploids. Moreover, cytoplasmic genomes also may be important in the success of polyploids, because in certain combinations they may foster nuclear–cytoplasmic heterosis. Some evidence that the cytoplasmic factor could be important in the ascendance of polyploids is presented below.

Organelle genomes may have a substantial effect on the phenotype of polyploids as seen in *Tragopogon*. For example, when *T. dubius* is the maternal parent of the tetraploid *T. miscellus,* the latter produces flowering heads with long ray-florets, whereas when *T. pratensis* is the maternal parent, *T. miscellus* produces flowering heads with short ray-florets (Ownbey and McCollum, 1953; Soltis et al., 1995). The fitness consequences of this difference remain to be determined.

It would not be surprising to find that polyploids with the same nuclear composition, but different maternal genomes, are divergent in some respects, given what we know about the effects of maternal parentage on the properties of interspecific F_1 hybrids at the diploid level. Differences in morphological traits have been described in reciprocal hybrids of *Gossypium* (Meyer, 1965, 1971, 1972), *Streptocarpus* Lawrence (1958), and *Avena* (Robertson and Frey, 1984). The differences may be manifold as in reciprocal hybrids of *Gilia tenuifolia* and *G. exilis* (Grant, 1956). They are divergent in the color of the herbage, stem pubescence, the size and shape of the corolla lobes, and corolla color. These differences may have fitness consequences.

We also know that nuclear–cytoplasmic interactions also may substantially influence the performance of interspecific hybrids. These interactions affect photosynthetic, respiratory, and growth rates in *Triticum* and allied genera (Iwanaga et al., 1978) and in *Helianthus* (Jan, 1992).

OVERVIEW

That polyploidy is important in flowering plants has been appreciated for many decades. The question has been the percentage of species with polyploidy in their evolutionary history. Is the value 30–35% as Stebbins estimated in 1971, or 47% as Grant estimated in 1963, or 70% as Masterson estimated in 1994? The latter apparently is the most informed, given what we are learning about genome replication from dense genetic maps. Such putative diploids as maize ($n=10$) and cabbage ($n=9$) now seem to be polyploids, as may be *Arabidopsis thaliana*, with its very small genome size and chromosome number ($n=5$).

For the most part, polyploids have been classified on the basis of genomic similarity. If a polyploid has the same genome in multiple copies, Stebbins and others have deemed it an autopolyploid, whereas if it has two or distinctive genomes it has been deemed an allopolyploid. Alternatively, Grant and others proposed that autopolyploids arise within populations, whereas allopolyploids are the products of crossing between species.

The production of unreduced gametes is a prerequisite for the formation of polyploids. The occurrence of unreduced eggs and sperm ranges widely within and among species, and in part is governed by genetic factors. These factors can lead to a wide variety of meiotic abnormalities.

Polyploids arise through one of two pathways. One involves the fusion of two unreduced gametes. The other involves the fusion of an unreduced gamete with a reduced gamete to yield a triploid, that in turn produces diploid or triploid gametes that subsequently combine with the reduced gametes of diploids. Both pathways are similarly plausible. The one taken by any given polyploid species cannot be determined.

The rate of polyploid species formation by one or the other path is a matter of conjecture, as is the rate of spontaneous genesis of polyploids within populations. The latter is more tractable, because it is subject to observation if one is willing to count the chromosome numbers of massive numbers of plants. Not many have been so willing, so we are left with a shallow database. Even diploid species prone to generating polyploids apparently do so less than one plant in five hundred.

The production of polyploids occurs within diploid populations, and accordingly polyploids initially will constitute a very small minority of the population. They face a vast hurdle in locally replacing their diploid counterparts, because most pollen they receive will be from diploids. If, for example, tetraploids are receptive to such pollen and flower synchronously with diploids, most their progeny will be triploid. The conditions required for the establishment of polyploids in the presence of diploids recently has been a subject of considerable interest. The replacement of diploids is not a prerequisite for polyploid establishment, which may ensue if their seeds reach suitable sites where diploids do not occur. Indeed establishment away from progenitors is probably the case for allopolyploids, which have unique tolerances due to chromosome doubling and hybridity. Polyploids usually replace their diploid prototypes along ecological gradients, suggesting that polyploids most often spread from the periphery of their prototype's niche space.

The ecological contrasts between diploids and polyploids frequently involve differences in temperature, light, and moisture regimes. However, the nature of diploid-polyploid divergence in one species assemblage has little predictive value for other assemblages. Interploidal differences in ecological tolerances usually are translated into different, although not necessarily mutually exclusive, geographical ranges. Interploidal differences also may be translated into different range sizes.

Transgeneric patterns in the ecological and geographical distributions of diploids and polyploids have been sought for several decades. We find that there is a tendency for polyploids to increase in frequency with increasing latitude. This does not necessarily mean that polyploids are more cold tolerant than diploids. Within regions, diploids and polyploids tend to have similar range sizes and range of habitat tolerances. Polyploidy is considerably more common in herbs than in wood plants.

Why are many allopolyploids so successful? Fixed heterozygosity long has been the explanation of choice. Another explanation is genetic novelty, which can be garnered through transgressive segregation or simply through segregation and recombination of parental genes prior to or after chromosome doubling. Gene interactions between disparate genomes also could be a contributory factor. Finally, there is the matter of nuclear-cytoplasmic interaction. Some hybrid nuclear backgrounds may be especially adaptive when embedded in the cytoplasm of one of the diploid parents.

7

Phenotypic Consequences
of Chromosome Doubling

It is important to uncouple the effects of chromosome doubling from those of hybridization or genome coexistence, because only then can we understand the manifold effects of polyploidy alone (autopolyploidy) within individuals, populations, and species. The effects of chromosome doubling per se generally have been viewed as maladaptive. Levan (1945) argued that chromosome doubling radically alters the balance of multiple factors acting on a single character and that only through hybridization and selection may this imbalance be restored. Stebbins (1971, p. 126) suggested that chromosome doubling alone would hinder the evolutionary success of higher plants. I take a contrary position. In this chapter I contend that autopolyploidy may generate novel types of plants and in turn may facilitate character evolution and speciation. In many species autopolyploidy may significantly alter the cytological, biochemical, physiological, and developmental attributes of organisms. Autopolyploids thus may have unique or transgressive tolerances and traits that may allow the exploitation of habitats beyond the limits of their diploid progenitors. Chromosome doubling occurs in existing polyploids, but it is more likely to produce a positive effect from a diploid base (Gottschalk, 1976). Thus, most comparisons that follow involve diploids.

CYTOLOGY

The most immediate and widespread effect of polyploidy in plants is an increase in cell size (Stebbins, 1971). The volume of tetraploid cells typically is about twice that of their diploid counterparts, whereas the surface area of tetraploid cells is about 1.58 times greater. The ensuing reduction in surface-to-volume ratio may cause a reduction in the rate of processes whose initiation and performance depend on both surface membranes and organelles.

The relationships between enzyme activity, cell geometry, and ploidal level have been studied in *Saccharomyces cerevisiae*. Weiss et al. (1975) found that the basic biochemical parameters of the cell are determined principally by cell geometry. Invertase, ornitine transcarbamalase, and tryptophan synthetase activities are related to cell volume, whereas the activity of acid phosphatase, a cell surface enzyme, is a function of the surface area of cells. It remains to be determined whether the same is true for higher plants.

Cavalier-Smith (1978) suggested that metabolism and growth would be retarded in polyploid cells because the geometric relationship between the nucleus and the remainder of the cell would be altered. RNA passes from the nucleus to the cytoplasm via nuclear pores, whose number per unit area is fairly constant between ploidal levels of the same species and even different species (Maul, 1977). For example, diploid *Hordeum vulgare* has 16.3 pores/μm^2, compared with 17.1 in the tetraploid. The rate of RNA transfer to the cytoplasm is dependent upon the area of the nuclear membrane, because the number of pores per unit area is relatively constant. Tetraploids have fewer pores per unit volume of cell, because the nuclear membrane of tetraploids will have about 1.58 times the area of the membrane of their diploid counterparts, whereas the volume of tetraploid cells is twice that of their diploid counterparts.

Polyploidy also lowers the ratio of the nuclear membrane to chromatin volume. As a consequence, more chromatin is in contact with the nuclear membrane, and a greater proportion of the chromatin is in condensed regions along the nuclear membrane. McClintock (1967) suggested that changes in chromatin–membrane relationships can alter gene expression and regulation.

GENE ACTIVITY

Polyploidy has the potential to affect the amplification of RNA cistrons and enzyme activity per unit protein. Consider first the amplification of RNA cistrons. Cullis and Davies (1974) reported an absence of differential amplification of ribosomal RNA in a ploidal series of *n, 2n, 3n, 4n,* and *6n* in *Datura innoxia.* Similarly, in *2n, 3n,* and *4n* varieties of *Hyacinthus orientalis,* RNA content per haploid genome was the same (Ingle et al., 1976). Even when doses of nucleolar organizing regions were varied through aneuploidy, product level per genome remained constant, thus indicating that RNA content was not regulated by specific gene dosage. In contrast, Tal (1977) found that autotetraploid tomato had about 15% less RNA than the diploid, and its RNAse activity was about 25% less.

Several investigators have shown that chromosome doubling usually impacts enzyme activity per unit protein. One of the most comprehensive studies was performed on *Lycopersicon esculentum* by Albuzio, Spettoli, and Cacco (1978). Regarding photosynthetic enzymes, they found that autotetraploids have about 10% less ribulose-1,5-bisphosphate carboxylase (RuBPC) activity and about 30% less phosphoenolpyruvate activity than do comparable diploids. The activities of ribulose-1,5-diphosphate oxygenase and glycolate oxygenase (both involved in photorespiration) also are affected by ploidal change. They are reduced by about 20% and 10%, respectively, in the autotetraploids. Regarding carbon metabolism, chromosome doubling nearly doubled the activities of malate dehydrogenase and acid invertase, and acid phosphatase activity increased about 25%. Neutral phosphatase and esterase activities were unaffected, but peroxidase activity declined about 50%. The activities of nitrate reductase and glutamate dehydrogenase, key enzymes of nitrogen metabolism, increased about 25% and 60%, respectively. Clearly, all enzymes are not affected in the same way by chromosome doubling. That may lead to a new, perhaps favorable, balance of metabolic regulation.

Guo, Davis, and Birchler (1996) investigated dosage effects in a maize ploidy series (n, 2n, 3n, 4n) in which rRNA per genome was approximately equal. They analyzed the expression of 18 genes and found that, for the most part, the absolute level of gene expression per cell increased with increasing ploidal level. However, on a per genome basis, there was approximately equal expression.

In *Arabidopsis thaliana*, a single-copy transgene for antibiotic resistance was introduced into diploid and triploid plants (Mittelsten Scheid et al., 1996). Gene activity was less in the triploid than in the diploid.

The affect of chromosome doubling depends on the genetic constitution of the organism. This is evident in differences among strains or populations of the same species. Dunbier et al. (1975) analyzed glucose-6-phosphate activity in three strains of *Medicago sativa* and their induced autotetraploids. Tetraploids of one strain had nearly twice the enzyme activity of it diploid counterpart, whereas in the other two strains enzyme activity was similar across ploidal levels.

Genotype-dependent effects of chromosome doubling also is seen in *Phlox drummondii*, where seven diploid populations and their synthetic autotetraploid counterparts were analyzed for alcohol dehydrogenase activity (Levin et al., 1978). In four populations chromosome doubling was accompanied by a doubling of enzyme activity, in two populations there was a 1.5-fold increase, and in one there was a small decline.

Not surprisingly, the effect of chromosome doubling may vary among species. Nakai (1977) studied esterase activity in diploids and autotetraploids in several genera. Increased activity accompanied chromosome doubling in 17 of 29 diploid-autotetraploid pairs in *Oryza*, two pairs in *Citrullus* and *Brassica*, and one pair each in *Astragalus, Beta, Ipomoea, Raphanus,* and *Vicia.* No difference (dosage compensation) was observed in four pairs of *Aegilops* and in two pairs of *Triticum.*

All of the studies referred to thus far have dealt with the effect of ploidal level on nuclear gene activity or gene activity in the chloroplast. The effect of ploidal level on mitochondrial gene activity has not been described. However, some insights into the matter might be gained when parts of the genome are multiplied. Such a study was conducted by Auger et al. (2001) on maize through the creation of a dosage series for 14 different chromosome arms. These arms were present one, two, or three times in a given plant. The expression levels of five mitochondrial genes were assayed. There was not a simple relationship between arm dosage and mitochondrial gene activity. It is noteworthy that the predominant effect of adding or subtracting arms was to reduce gene expression. This may explain the observation that the plants with deviant arm numbers are less vigorous than their siblings with the normal arm number. Whether or not haploid or triploid maize show the same retardation of mitochondrial gene expression remains to be determined.

PHYSIOLOGY

Polyploidy affects the hormone status, water relationships, photosynthetic rate, and secondary metabolism of plants, as discussed below.

Hormones

The effect of chromosome doubling on growth hormones is an important issue, because chromosomes have an important regulatory influence on most plant functions. Unfortunately, rather little is known about this matter. Avery and Pottorf (1945) found two to three times as much auxin in diploid leaves and stems of cabbage as in the tetraploids. Abscisic acid content was about 30% lower in autotetraploid tomatoes than in the diploid (Tal, 1980). In a naturally occurring series of ploidal levels in *Sedum pulchellum,* diploids had the highest hormone levels, tetraploids somewhat less, and hexaploids the least (Smith, 1946). In no ploidal series has there been a simultaneous analysis of all of the major hormones and hormone antagonists. Because hormones usually interact with each other, such analyses would be particularly informative.

Water relations

Ploidal races within species typically differ in their habitat preferences, as discussed below. Thus, it is not surprising that chromosome doubling can have a substantial impact on plants' water relations. Little is known about differences between diploids and polyploids in water uptake, but transpiration has been studied in a few species. Given that the stomatal apparatus controls transpiration, it is noteworthy that chromosome doubling typically is associated with a decline in stomatal density (e.g., *Ribes satigrum,* Cukrova and Avratobscukova, 1968; *Triticosecale,* Sapra et al., 1975; *Medicago sativa,* Setter et al., 1978; *Betula papyrifera,* Li et al., 1996). In these species and most others, this decline is accompanied by an increase in stomatal size. However, the increase in size usually does not counterbalance the decline in number, because the stomatal pore space per unit area is less in polyploids. Accordingly, chromosome doubling is likely to increase does leaf diffusive resistance. Because transpiration rates are a negative function of diffusive resistance, they are likely to be lower in polyploids than in diploids.

One of the more illuminating studies on water relations and polyploidy was conducted on diploid and pentaploid *Betula papyrifera.* Li, Berlyn, and Ashton (1996) reported that pentaploids have a significantly lower water potential at saturation than diploids, −1.61 versus −1.50. This suggests that the pentaploids are more able to maintain turgor during low tissue water potential than are the diploids. Under artificial water stress, the diploids exhibited a greater decline in net photosynthesis and stomatal conductance than did the pentaploids. The differences between the ploidal levels arise in part from the thicker upper and lower epidermis of the polyploids and their fewer stomata per unit area. It is not surprising that the diploids favor wetter soil environments than the pentaploids.

Lower transpiration rates also have been reported in polyploid barley (Chen and Tang, 1945) and tomato (Tal, 1980). In tomato the rate in the tetraploid was 84% that of the diploid. Lower rates, however, are not an inevitable outcome of chromosome doubling as seen in *Medicago sativa,* even though polyploids have lower stomatal pore space per unit area (Setter et al., 1978).

Photosynthesis

Consider next the CO_2 exchange rate, which measures the effectiveness of a plant's integration of many biochemical processes. Given that leaf diffusive resistance is inversely correlated with photosynthetic ability (Wong et al., 1979; Byrne et al., 1981), CO_2 exchange rates in polyploids are expected to be lower than in their diploid progenitors.

Until the 1970s, photosynthetic rates typically were reported on a per unit leaf basis. Recent studies report rates on a per cell basis as well as per leaf area basis. The photosynthetic rate per unit leaf area is the product of the rate per cell times the number of photosynthetic cells per unit area (Warner and Edwards, 1989). Accordingly, the photosynthetic rate per unit leaf area is expected to increase as ploidal level increases if there is less than a proportional increase in cell volume as ploidal level increases, or if cell packing is altered to allow more cells per unit leaf area.

On a unit leaf area basis, the photosynthetic rate of tetraploids averages about 10% less than diploids (e.g., *Ribes satigrum,* Bjurman, 1959; *Datura stramonium,* Cukrova and Avratocscukova, 1968; *Raphanus sativus,* Frydrych, 1965; *Hordeum vulgare,* Ekdahl, 1944; *Galeopsis pubescens,* Ekdahl, 1949; *Phlox drummondii,* Bazzaz et al., 1983). Similarly, photosynthetic rates are about 15% lower in hexaploid *Panicum virgatum* than in its tetraploid counterpart (Wullschleger et al., 1996). No response to ploidal level change was reported in *Medicago sativa* (Setter et al., 1978), *Dactylis glomerata* (Eagles and Othman, 1978), or *Viola adunca* (Mauer et al., 1978). Then there is the converse of the rule in *Beta vulgaris,* where tetraploids had a higher CO_2 fixation rate than diploids (Beysel, 1957).

Warner and Edwards (1989) studied a series of *Atriplex confertiflora* races ranging from diploid ($2x$) to decaploid ($10x$). The polyploid races occupy the sites of former Pleistocene lakes, and thus ostensibly arose within the past 10–15 thousand years. There was a significant positive correlation between ploidal level and photosynthetic rate on a per cell basis, and significant correlations between ploidal level and activities of ribulose-1,5-bisphosphate carboxylase per cell and NAD-malic enzyme per cell (figure 7.1). The relationship between gene dosage and gene products was nearly proportional, as was the amount of chlorophyll in bundle sheath cells. Conversely, the amount of chlo-

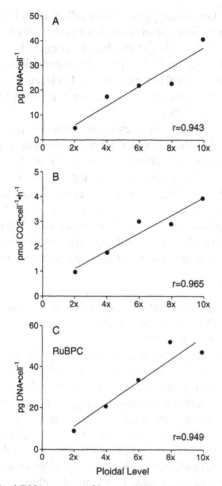

Figure 7.1.   (A) Total DNA content of leaves expressed per photosynthetic cell in *Atriplex confertifolia.* (B) Photosynthetic rates per cell in leaves. (C) RuBPC activity per bundle sheath as functions of ploidal level. Redrawn from Warner and Edwards (1989), with permission of the American Society of Plant Physiology.

rophyll per mesophyll cell did not change with ploidal level. Photosynthetic rates per unit leaf area were relatively constant at the lower three ploidal levels. This occurs because cell number per leaf area decreases as the cell size and photosynthetic capacity per cell increase. Photosynthesis per area was greater in octoploids and decaploids than in the other ploidal levels.

An autotetraploid of pearl millet (*Pennisetum americanum*) had twice the photosynthetic rate per cell of its recent diploid progenitor (Warner and Edwards, 1988). There are half as many cells per unit leaf area in tetraploids, so they have the same photosynthetic rate per unit area as diploids. In *Panicum virgatum,* an octoploid strain has twice the photosynthetic rate per cell as a tetraploid strain

(Warner et al., 1987). Given that the number of cells per unit leaf area decreases somewhat more then the increase in cell volume in the octoploid, the latter has a photosynthetic rate that is 40% greater than the tetraploid.

In *Festuca arundinacea,* the CO_2 exchange rate per unit leaf area increases as we progress from the 4x level to the 6x, 8x, and 10x levels (Byrne et al., 1981). This increase is related to a concomitant increase in ribulose-1,5-bisphosphate carboxylase (Joseph et al., 1981).

Thus far, consideration has been given to the effects of chromosome doubling independent of hybridization. However, most polyploid taxa have both chromosome doubling and hybridization in their evolutionary histories. Thus, it is informative to see the consequences of doubling the chromosome number in hybrids. Hiesey, Nobs, and Björkman (1971) reported that between 10°C and 25°C, diploid F_1 hybrids between *Mimulus cardinalis* and *M. lewisii* had a considerably higher photosynthetic rate than either parent. It is noteworthy that the tetraploid derivative of the F_1 hybrid and tetraploid derivative of an F_2 hybrid had higher photosynthetic rates than their diploid counterparts.

Secondary metabolism

Chromosome doubling often is accompanied by conspicuous changes in secondary metabolism. These changes may alter the interactions between plants and their herbivores and pathogens.

Autopolyploids in many drug plants including species of *Vinca, Lobelia, Atropa,* and *Rauwolfia* have sharply increased quantities of alkaloids per unit weight (Banerjee, 1968; Dnyansager and Sudakaran, 1970). The alkaloid content in *Datura tatula* tetraploids was twice as high as in diploids (Rudolph and Schwartz, 1951). In *Papaver bracteatum* fruits, thebaine content was twice as high in tetraploids as in diploids (Milo et al., 1987), and in fruits of *Solanum khasianum,* solasodine content was 35–50% higher in tetraploids (Bhatt and Heble, 1978). The tropane alkaloid yield in *Atropa belladonna* was 68% higher in tetraploids than in diploids (Evans, 1989), and this class of compounds was 22% higher in *Hyocyamus niger* tetraploids (Lavania and Srivastava, 1991) and 36% higher in *H. muticus* tetraploids (Lavania, 1986).

Terpene contents also tend to be higher in autopolyploids than in diploids. In *Carum cari,* chromosome doubling increased terpene content from 35% to 85%, depending on the strain and location (Zderiewicz, 1971; Dijkstra and Speckmann, 1980). In *Mentha arvensis,* terpene content in the polyploid was 30% higher than in the diploid (Janaki Ammal and Sobti, 1962), and in *Ocimum kilmandscharicum,* it was 50% higher (Bose and Choudhury, 1962). As a result of chromosome doubling, volatile oil content increased 300% in *Acorus calamus* (Evans, 1989), and essential oil content increased 60% in *Vetiveria zizanoides* (Lavania, 1988).

The effect of chromosome doubling within a species may depend on the genetic background. For example, coumarin levels in autotetraploid *Artemisia tridentata* ssp. *tridentata* are twice as high as in the diploid, whereas in autotetraploids of ssp. *vaseyana,* coumarin levels are about 45% less than in diploids (McArthur and Sanderson, 1999).

Chromosome doubling also may alter the secondary chemical profile of a plant in a qualitative manner. Levy (1976) found such differences in the gly-coflavone profiles of 14 of 15 synthetic autotetraploid *Phlox drummondii* populations as compared to their diploid prototypes. This included 14 instances of flavonoids present in the polyploid but absent in the diploid, and eight instances of the flavonoids present in the diploid but absent in the polyploid. Qualitative differences between ploidal levels also have been found in *Rubus* (Haskell, 1968) and *Briza* (Murray and Williams, 1976).

Griesbach and Kamo (1996) analyzed the effects of ploidal level change on the proportional balance among quercetin compounds in the *Petunia* cultivar Mitchell. The most pronounced change was in quercetin-3,7-diglucoside, which declined from 14% in haploids to 8.5% in diploids to 3.5% in tetraploids. At the same time quercetin-3-glucoside increased from 73.6% to 79.4% to 81.3%, respectively.

DEVELOPMENT

Autotetraploids typically have larger cell sizes and lower growth rates than related diploids (Gottschalk, 1976). These differences are expressed variously throughout the life cycle from the size and germination schedules of seeds to plant longevity.

The seeds of most autotetraploids are larger than their diploid counterparts, sometimes more than twice as large (Gottschalk, 1976). Autotetraploid seed typically germinate slower than diploid seed. For example, autotetraploid seed of *Lycopersicon esculentum* germinates 3–5 days later than diploid seed (Jorgensen, 1928).

Autotetraploid seed tends to have a lower germination percentage than diploid seed (Gottschalk, 1976). In some instances, this represents a dormancy differential rather than a viability differential (e.g., *Corchorus olitorius* and *C. capsularis,* Datta, 1963).

The large size of tetraploid seeds often results in their seedlings being more robust and faster growing than those of the diploids, as demonstrated in muskmelons (Batra, 1952) and subterranean clover (Hutton and Peak, 1954). Even the development of the embryo and endosperm of tetraploids may be more rapid (Pandey, 1955). However, when polyploids display greater vigor during seed and seedling development, it usually does not carry over to the adult stage of the life cycle.

Garbutt and Bazzaz (1983) studied the leaf demography and flower production in three diploid–tetraploid pairs of *Phlox drummondii* growing on a soil moisture gradient. Tetraploids tended to grow slower in the drier portions of the gradient, as evident in their lower biomasses at harvest and lower intrinsic rates of leaf population growth (figure 7.2). Polyploids tended to produce fewer but somewhat larger flowers than do diploids.

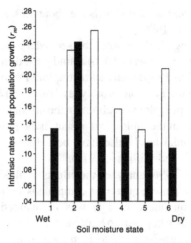

Figure 7.2.   Intrinsic rates of leaf population growth in each of six moisture states for diploid (open bars) and tetraploid (solid bars) *Phlox drummondii* ssp. *mcallisteri*. Redrawn from Garbutt and Bazzaz (1983), with permission of *New Phytologist*.

Autopolyploids typically grow slower than diploids, and this may result in the former flowering later or over a longer time span. For example, autotetraploid *Ocimum kilimandscharium* flowers 30–45 days later than its diploid counterpart, with the duration of the former being about a month and a half longer than that of the latter (Bose and Choudhury, 1962). A similar differential was observed between 2n and 4n tomato (Kostoff and Kendall, 1934). Tetraploid *Phlox drummondii* reaches its flowering peak about 30 days later than its diploid progenitor and remains in flower for as much as a month after the diploid has ceased (Garbutt and Bazzaz, 1983).

The disparate growth properties of diploids and autotetraploids may be reflected in their longevities, including a shift from the annual to perennial habit. In *Oryza punctata*, tetraploids survive for 2–9 years, whereas diploids are almost always annuals (Sano, 1980). Shifts from annual to perennial habit also have been described in *Zea mays* (Müntzing, 1936), *Eragrostis* (Hagerup, 1932), and *Nasturtium* (Manton, 1935). In garden trials, *Ocimum kilimandscharium* diploids have a mean life span of about 2 years, whereas tetraploids live for about 3 years (Bose and Choudhury, 1962). Greater longevity also has been reported in tetraploid *Trifolium pratense* (Frame, 1976) and *Medicago lupulina* (Schröck, 1944).

REPRODUCTIVE SYSTEM

A change in ploidal level may have a substantial impact on the reproductive biology of a species. A breakdown in self-incompatibility is the most common alteration, and this occurs in species with single-locus gametophytic self-incompatibility (e.g., *Lycopersicon peruvianum*, de Nettancourt et al., 1974; *Trifolium hybridum*, Brewbaker, 1958). The breakdown does not arise from the failure of the pistil to recognize pollen sharing S-alleles but rather from the loss of specificity in the diploid pollen grains (Stout and Chandler, 1942; Crane and Lewis, 1942). However, not all diploid pollen is compatible. In some species with single-locus gametophytic self-incompatibility, it appears that only grains carrying two different S-alleles are compatible (de Nettancourt, 2001). In other species, it also depends on the genetic background of the pollen producers.

Polyploidy does not foster self-compatibility in species with multilocus complementary gametophytic incompatibility. Nor does it foster self-compatibility in species with sporophytic incompatibility (de Nettancourt, 2001).

The penchant for selfing may be strengthened in polyploids by alterations in floral architecture. For example, in *Lamium amplexicaule* chromosome doubling increases the percentage of cleistogamic flowers (Bernstrom, 1950). Chasmogamic flowers are self-compatible but produce fewer seeds per flower than do the diminutive cleistogamic flowers.

Chromosome doubling may alter the reward of individual flowers. Maurizio (1954) compared the nectar production of individual flowers per unit time in diploid and polyploid strains of five *Trifolium* species, two *Salvia* species, two *Datura* species, and *Lobelia syphilitica*. Nectar production was 1.5–4.5 times greater in the polyploids. In diploid, triploid, tetraploid, and octoploid races of *Salvia splendens,* nectar production per flower per day increased with ploidal level.

Through its effect on floral form and floral attractants, polyploidy may alter the relationship between plants and their pollinators. For example, in *Trifolium repens,* tetraploids have longer and wider corolla tubes than do diploids (Taylor and Smith, 1979). The longer tubes makes pollination by honeybees and other insects more difficult, which renders the diploids the preferred entity.

The most comprehensive study of pollinator visitation to mixed populations of diploids and autotetraploids was conducted on *Heuchera grossularifolia* along the Salmon River in Idaho (Segraves and Thompson, 1999). The tetraploids had longer petals, deeper corolla tubes, and fewer flowers per inflorescence than did the diploids. The simultaneously flowering cytotypes received similar number of visits per unit time. However, the assemblage of pollinators visiting the cytotypes differed significantly. The bees *Osmia kincaidii, Halictus farinosa,* and species of *Lasioglossum* and *Nomada* visited diploid plants more than twice as often as they visited tetraploid plants. Conversely, *Bombus* tended to prefer the tetraploids, and the lepidopteran *Greya politella* and the dipteran *Bombyllius major* visited tetraploids at least five times more frequently than diploids (see table 6.2).

An increase in ploidal level alone may cause a shift from sexual reproduction in diploids to partial or complete apomixis in polyploids. The effect of ploidal changes alone is very well documented in the grass genus *Paspalum.* Sexual diploid races and predominantly apomictic tetraploid races occur in *P. brunneum, P. haumani, P. intermedium, P. maculosum,* and *P. rufum* (Norrmann et al., 1989). Diploids develop an aposporous embryo sac in addition to a meiotic embryo sac, thus demonstrating their potential for apomixis. Of particular importance in showing cause and effect is the fact that induced tetraploids in *P. rufum* (Quarin et al., 1998), and induced tetraploids and hexaploids in *P. hexastachyum* are predominantly apomictic, whereas their diploid counterparts are completely sexual. A rise in ploidal level also induces apomixis in *P. notatum* (Quarin et al., 2001) and *P. compressifolium* (Quarin et al., 1996). Chromosome doubling in other grasses such as *Calamagrostis canescens* (Nygren, 1948) and *Brachiaria decumbens* (Naumova et al., 1999) also is accompanied by a shift toward apomixis. Given the effect of polyploidy on the reproductive systems of the aforementioned grasses, it is not surprising that all apomictic species are polyploids (Savidan, 2000).

An increase in genome number is no guarantee of a change in reproductive mode, even if apomixis is present at higher ploidal levels. For example, in *Parthenium argentatum,* autotetraploids obtained from sexual diploids are completely sexual (Hashemi et al., 1989). Synthetic autotetraploids of sexual diploid *Taraxacum kok-saghyz* are completely sexual, although all triploid and tetraploid congeners in the same region are apomictic (Stebbins, 1950).

Increasing the ploidal level may even increase the level of sexual reproduction. This result was reported in *Poa pratensis* (Müntzing, 1940) and *P. palustris* (Zirov, 1967, cited in Asker and Jerling, 1992), where triploids had a higher degree of sexuality than did their diploid counterparts. In *Paspalum longiflorum,* octoploids produced by the colchicine treatment of an obligately apomictic tetraploid had normal meiosis (Chao, 1974).

Whereas polyploidy per se may have an immediate impact on a species' reproductive biology, it also may set the stage for a shift from cosexual (perfect) flowers to gender dimorphism in the form of dioecy and gynodioecy. Miller and Venable (2000) present evidence for this shift in 12 genera involving at least 12 independent evolutionary transitions. They propose that polyploidy disrupts self-incompatibility, thus leading to self-compatible polyploids that suffer from inbreeding depression. This in turn favors the invasion of male sterility and ultimately a stabilized system of dioecy or gynodioecy. The pivotal point in this scenario is inbreeding depression in polyploids, the magnitude of which is poorly understood.

ECOLOGICAL TOLERANCE

We have seen that a shift in ploidal level may alter some fundamental properties of plants. These shifts and others may affect the ecological tolerances of species by altering their habitat tolerances and their interactions with competitors and with pests and pathogens.

Nutrient tolerance

Some polyploids may have the ability to occupy nutrient-deficient soils that their diploid counterparts can not. In *Dianthus* (Rohweder, 1937) and *Nicotiana* (Noguti et al., 1940), tetraploids thrive on calcareous soils, whereas diploids cannot grow there. Greater tolerance to lower nutrient levels of these polyploids and polyploids in other genera may arise from their superior ion uptake efficiency. One of the more comprehensive studies on this issue was conducted by Cacco, Ferrari, and Lucci (1976). They compared the uptake efficiency of K^+ and SO^{-2} in diploids and autotetraploids of *Lycopersicon esculentum* and diploids, triploids and tetraploids in *Beta vulgaris*. The uptake efficiency for the cation and anion increased with ploidal level in *Beta*, but declined in *Lycopersicon*.

The relative uptake efficiencies of plants with different ploidal levels may be reflected in their mineral contents. The calcium, magnesium, potassium, phosphorus, and sulfur content in tetraploid tobacco (Noguti et al., 1940) and potassium and phosphorus levels in tetraploid *Galeopsis pubescens* (Ekdahl, 1949) are greater than in their diploid counterparts. The ash contents of autotetraploids in many genera are higher than the contents of their diploid prototypes (*Hordeum, Nicotiana, Petunia, Trifolium;* Noggle, 1946). The allotetraploid derivative of the diploid F_1 *Nicotiana rustica* × *N. paniculata* has an ash content that is 50% greater than of the diploid.

There is no evidence that the uptake efficiency of nitrogen is greater in polyploids (Noggle, 1946). However, in poor soil autotetraploid *Medicago sativa* has higher nitrogen levels than the diploid. This is because the tetraploid has a higher rate of nitrogen fixation in the root nodules (Leps et al., 1980), a feature that may explain why the tetraploid consistently has higher yields (Dunbier et al., 1975; Arbi et al., 1978). Tetraploid *Trifolium pratense* and other clovers also are consistently superior to their diploid counterparts (Frame, 1976).

Soil moisture tolerance

There is a moderate amount of anecdotal evidence that some polyploids are more tolerant of drought than are their diploid prototypes. The best evidence comes from the differential performance of diploids and tetraploids over a series of environments. The divergent habitat tolerances of some diploids and polyploids also point to differences in drought tolerances, as discussed in chapter 6.

Beginning with greenhouse studies, Garbutt and Bazzaz (1983) determined the shoot biomass for three diploid cultivars of *Phlox drummondii* and their synthetic autotetraploids for plants growing along a six-step soil moisture gradient. The tetraploids performed best in drier states than their diploid prototypes. Autotetraploids of *Caragana arborescens* also were more drought tolerant than their diploid progenitors (Pustovoitova and Borodina, 1981). Under drought conditions, the tetraploids had higher water-retaining capacity than the diploids.

To compare habitat tolerances of diploid and tetraploid cytotypes of the grass *Ehrharta erecta*, Stebbins (1949) planted seeds of both cytotypes in a series of natural sites near Berkeley and Carmel, California. After a few years it became evident that the tetraploids preferred well-drained soils, whereas the diploids preferred more mesic sites. The same preference differential emerged in a second planting of the same type (Stebbins, 1972). Tetraploid preference for more xeric sites also was evident in $2n$ and $4n$ mixtures of *Arabidopsis thaliana* planted in a series of field sites (Bouharmont and Mace, 1972).

Temperature tolerance

Chromosome doubling may be accompanied by increased resistance to cold temperatures. For example, autotetraploids are more resistant to cold in *Brassica campestris* (Choudhury et al., 1968), *Raphanus sativa* (Nishiyama, 1942), *Lolium multiflorum* (Wit, 1958), *Melilotus officinalis* (Schröck, 1944), and *Trifolium pratense* (Goral et al., 1964). Conversely, tetraploids are less tolerant of cold in *Trifolium repens* (Goral et al., 1964), *Arrenatherium elatius* (Zimmermann, 1968), *Secale cereale* (Dvorák and Fowler, 1978), and naturally occurring *Festuca pratensis* (Tyler et al., 1978).

Hall (1972) reported that tetraploid rye is less tolerant of high soil temperatures than are its diploid counterparts. He reasoned that low oxygen tension in warm soils may prevent the spread of the tetraploids and contribute to the different distributions of the diploids and tetraploids.

Chromosome doubling may alter the temperature optima for various physiological processes, the most notable being photosynthesis. In *Trifolium incarnatum*, the temperature optimum for diploids at high light intensity was 15–20°C versus 10°C for tetraploids (Wohrmann and Drew, 1959). Induced tetraploids of *Dactylis glomerata* had higher rates of photosynthesis at lower temperatures and developed more rapidly than their diploid prototypes (Eagles and Othman, 1978).

Resistance to pests and pathogens

Greater resistance to pests and pathogens often accompanies a ploidal increase. This may be due to an increase in levels of secondary products or to an alteration in some physiological plant properties.

Burdon and Marshall (1981) infected diploid and autotetraploid populations of *Glycine tabacina* with the leaf-rust fungus *Phakospora pachyrhizi*. Forty-two percent of the tetraploids were resistant versus 14% of the diploids. Among the susceptible plants, the rate of disease development and the number of pustules per unit area of leaf were lower in the tetraploids. A greater resistance to the clover rot fungus *Sclerotinia trifoliorum* was found in tetraploid *Trifolium pratense* than in diploids (Vestad, 1960). Isogenic lines differed in survival after 70 days, the mean for the diploid being 54% versus 67% for the tetraploid. Tetraploid *Raphanus sativa* was more resistant to club foot root disease than was the diploid (Nishiyama, 1942), and tetraploid *Beta vulgaris* was more resistant to *Cerospora* (Gottschalk, 1976). Chromosome doubling imparted greater resistance to dieback and collar- and root-rot diseases in *Catharanthus roseus* (Kulkarni and Ravindra, 1988).

Herbivory also may be affected by chromosome doubling. The stinging nematode reduced the dry weight of St. Augustine grass roots by 33% relative to controls, whereas polyploids were unaffected (Busey et al., 1993). Tetraploid *Trifolium pratense* had greater resistance to the clover eel nematode and to insects than the diploid (Mehta and Swaminathan, 1957). Tetraploid *Brassica campestris* was less affected by aphids than were diploids (Choudhury et al., 1968).

Thompson et al. (1997) proposed that chromosome doubling may create evolutionary barriers to attack by the pests that plague diploids. They tested this idea by analyzing the colonization success of the specialist moth herbivore *Greya politella* on the saxifrage *Heuchera grossularifolia*. Contrary to expectations, they found that the moth is more likely to attack the tetraploids in populations where the cytotypes coexist. Nevertheless, the idea has merit and may be operative in other situations.

COMPETITIVE ABILITY

Two types of experiments have been conducted to assess the competitive abilities of diploid and tetraploids. In one type, plants are grown in pots or in the field in pure and mixed cultures. Then the biomasses of both cytotypes are determined at the end of the growing season as are deviations in mixtures from pure culture. In the other experiment, seeds of the two cytotypes are sown in equal numbers and the numbers of each cytotype are followed across generations.

The few studies of performance in mixtures show that ploidal level indeed affects the outcome of competition. For example, Sakai (1956) planted diploid and tetraploid cytotypes of the tobacco cultivars White Burley and Bright Yellow in pure stand plots and in plots where cytotypes of a given cultivar alternated. Tetraploids were the inferior competitor in both cultivars, as seen their lower top weights and flower numbers. The suppression of tetraploids by diploids also has been documented in *Oryza sativa* (Sakai and Utiyamada, 1956) and *Hordeum vulgare* (Sakai and Sukuki, 1955a).

An example of the competitive superiority of tetraploids comes from *Trifolium pratense*. In mixed cultures of diploid and tetraploids planted as seedlings, the latter had a growth rate that was nearly twice that of the former (Anderson, 1971).

The competitive status of allotetraploids relative to the diploid species whose genomes they carried has been studied in *Nicotiana* and *Abelmoschus* (Sakai and Suzuki, 1955b). The tetraploids were superior in both genera. Unfortunately, the tetraploids did not compete with their immediate progenitors (F_1 hybrids), so we cannot assess the affect of chromosome doubling independent of hybridization. In *Oryza sativa*, tetraploids have lower competitive ability than the F_1 intervarietal hybrids from which they were derived (Sakai and Utiyamada, 1956).

The aforementioned experiments and others like them do not take seed germination into account. Thus, their value in predicting the course of change in mixed populations in the field is somewhat limited, because the percentage and timing of germination may differ among cytotypes. One informative cross-generation experiment involved the self-fertilizing *Arabidopsis thaliana*. Bouharmont and Mace (1972) sowed equal numbers of diploid and tetraploid seed in experimental plantings. Seeds were collected from mature plants and resown to establish the next generation. This procedure was conducted for five generations. The proportion of tetraploid plants increased for five successive generations. This was because tetraploids produced more seeds than diploids per plant, and seeds of the former had somewhat higher germination. In a 2-year study of competition between naturally occurring diploid and autotetraploid *Dactylis glomerata*, the latter increased in frequency (Maceira et al., 1993).

Diploids and polyploids do not necessarily differ in competitive ability. Similar competitive abilities were observed in mixtures of diploid and tetraploid rye over a 3-year period (Hagberg and Ellerström, 1959). The frequency of the cytotypes in mixtures remained roughly the same within growing seasons whether the cytotypes were initially at the same frequencies or whether one was in the clear majority (96%).

OVERVIEW

The phenotypic consequences of chromosome doubling independent of hybridity are manifold. Chromosome doubling increases cell size while reducing the rates of mitosis and meiosis. Several investigators report that enzyme activity per unit protein is affected, but not necessarily in the same direction. Furthermore, the level of change in enzyme activity is genotype-dependent. Transpiration and photosynthetic rates often differ among ploidal levels, but again the direction of change with chromosome doubling varies with the species. As these changes may cause a niche shift in polyploids, so may changes in mineral uptake efficiency and soil moisture tolerance. These changes are highly species-specific.

Many polyploids have slower growth than their diploid counterparts. In some instances, polyploids are perennial, while their progenitors are annual. Perennial polyploids sometimes have longer life spans than related diploid perennials. Changes in growth rate also may propel a species from one niche to another.

Given that autopolyploids initially occur in mixtures with their diploid ancestors, they must compete with them for a range of resources. Unfortunately, very little is known about the competitive status of polyploids, and this is not an area of current research. We can say that in some species autopolyploids are superior competitors, while in others they are inferior. Almost nothing is known about the competitive status of allopolyploids and their parental diploids.

Chromosome doubling often is accompanied by conspicuous changes in secondary metabolism. The alkaloid content of polyploids sometimes is 50% greater (or more) per unit weight than in diploids. Increases also have been described for other classes of compounds. Alterations in the chemical profile may accompany altered levels of compounds. Chemical changes associated with polyploidy have the potential to alter species' relationships with predators and pathogens, because many secondary chemicals have a role in plant defense. Indeed it is possible that polyploidy might allow species to escape from some of these organisms. Greater resistance has been reported in some polyploids, but whether it relates to alterations in secondary chemistry or to other plant attributes remains to be determined.

Alterations in plant-pollinator interactions and the breeding system also may be byproducts of chromosome doubling. A breakdown in self-incompatibility is the most common outcome in species with single-gene gametophytic incompatibility systems. Species with single-gene sporophytic incompatibility are unaffected as a species with multilocus systems. Chromosome doubling also may alter a flower's morphology and nectar production, thereby affording pollinators distinctive images that may promote assortative mating within ploidal levels. Changes most often are in the direction of larger flowers and increased nectar production. Polyploidy also is known to induce partial or complete apomixis. Indeed most apomicts are polyploid.

8

Chromosomal and Genetic
Consequences of Chromosome
Doubling

MEIOTIC CONSEQUENCES OF AUTOPOLYPLOIDY

Abnormal chromosome pairing

Autotetraploids have four homologues that are capable of pairing. Only two chromosomes may be paired at any one position. However, different chromosomes may be paired at different positions. As a consequence, meiocytes in prophase and metaphase I of meiosis may contain quadrivalents. The number of quadrivalents and other possible configurations (trivalents plus univalents and bivalents) depends on the number of chiasmata per chromosome (Sybenga, 1992). The greater the number, the more likely quadrivalents will be formed.

Several models of chromosome pairing behavior have been formulated for autotetraploids to estimate the frequencies of univalents, bivalents, and multivalents with random pairing among all chromosomes (Sybenga, 1975; Jackson and Casey, 1982; Jackson and Hauber, 1982). The general consensus is that if chromosome pairing is random, synapsis is initiated at both chromosome ends, and if chiasmata are located distally, then the frequency of pachytene multivalents will be two-thirds.

Synthetic and spontaneous polyploids have substantial levels of multivalent (trivalent and quadrivalent) formation. Examples of this phenomenon are presented in table 8.1. In *Triticum, Hyoscyamus,* and *Lathyrus,* most of the chromosomes are involved in multivalent configurations. The incidences of pairing configurations approach those predicted by the aforementioned models.

High frequencies of multivalents do not prevail in all synthetic autopolyploids. For example, the mean number of quadrivalents in *Chrysanthemum boreale* ($4x = 36$) was 1.29 and the mean number of bivalents was 15.14 (Watanabe, 1983).

Table 8.1. The occurrence of multivalents in autotetraploids

Species	2n	Multivalents	Bivalents	Univalents	Reference
Triticum speltoides	28	3.97	6.10	0.15	Yen and Kimber (1990)
Triticum bicorne	28	3.70	6.26	1.20	Yen and Kimber (1990)
Secale cereale	28	2.47	9.03	0.60	Moore (1963)
Hyoscyamus muticus	56	8.80	9.10	1.30	Srivastava et al. (1992)
Lathyrus odoratus	28	6.22	1.57		Khawaja et al. (1997)
Crepis capillaris	12	2.58	0.84		Jones and Vincent (1994)
Lotus alpinus	24	0.77	8.80	3.28	Somaroo and Grant (1971)
Lotus tenuis	24	1.20	8.97	5.60	Somaroo and Grant (1971)
Haplopappus spinulosus	16	0.68	13.28		Hauber (1986)

As noted in chapter 7, chromosome doubling may promote apomixis. The question thus arises as to the nature of chromosome pairing in polyploid apomicts, and whether sexual and apomict polyploids have different pairing relationships. This issue is best addressed in the grass genus *Themeda* (Birari, 1980). Diploid (sexual) species have 10 pairs of chromosomes during prophase I of meiosis, and sexual tetraploid species have 20 pairs of chromosomes. In contrast, partially and completely apomictic species at the tetraploid level and higher have multivalents and univalents in addition to bivalents (table 8.2). They have reduced gamete fertility, the disadvantage of which is overridden by their ability to produce progeny without gametes.

In general, do all chromosomes have equal probabilities of participating in multivalent pairing? Khawaja, Sybenga, and Ellis (1997) found that the probability of being involved in multivalent pairing does not differ significantly between chromosomes in induced autotetraploids of *Lathyrus odoratus, L. pratensis,* and *L. sativus.*

This pattern is not universal. Consider *Crepis capillaris,* which has a small genome composed of three distinctive chromosomes. The two longer chromosomes participated in quadrivalents in about 95% of the cells in prophase I, whereas the smaller one participated in only 68% of the cells (Jones and Vincent, 1994).

The failure of all chromosomes to have similar multivalent frequencies in

Table 8.2.   Chromosomal pairing associations at metaphase I in *Themeda*

Species	Chromosome number (2n)	Mean Chromosome configuration	Reproductive mode
T. anthera	20	10II	Sexual
T. tremula	20	10II	Sexual
T. strigosa	20	10II	Sexual
T. villosa	40	20II	Sexual
T. caudata	40	20II	Sexual
T. quadrivalis	40	3I + 13.5II + 0.7III + 2.00IV apomicts	Facultative
T. triandra	40	2.4I + 15.2II + 1.8IV apomicts	Facultative
T. triandra	80	0.4I + 30.4II + 2IV + 1.8VI apomicts	Obligate
T. dracruzii	60	16.4II + 2.0III + 2.1IV + 2.1IIV apomicts	Facultative
T. arundinacea	60	1.4I + 25.0II + 1.0III + 1.4IV apomicts	Facultative

Source: Adapted from Berari (1980).

Crepis capillaris and in other species is related to differences between them in chiasma frequencies. Similarly, differences in the mean multivalent frequencies of chromosomes between strains or species are related to differences in chiasma frequencies. The higher the mean chiasma frequency the more chromosomes are involved in multivalents (Khawaja et al., 1997).

Of particular interest is the fact that in many species there is a positive correlation between the chiasma frequency of diploids and their autotetraploid derivatives. One notable study is on *Hyoscyamus muticus,* where autotetraploidy was induced in 19 morphologically distinct genotypes (Srivastava et al., 1992). The multivalent association frequencies at metaphase varied from 0.33 to 0.63 and were positively correlated with chiasma frequencies in the diploid prototypes ($r = 0.65$; figure 8.1). Positive correlations between diploids and derived tetraploids have been described in several other species, including *Lolium perenne* (Simonsen, 1973), *Festuca pratensis* (Simonsen, 1975), and *Lathyrus sativus* (Khawaja et al., 1977). Such correlations are indicative of the genetic control of chiasma frequencies and thus chromosome associations.

We may be tempted to assume that because multivalent frequencies may be associated with chiasma frequencies, the mean frequencies of multivalents for species will be positively correlated with their mean chromosome sizes. This seems not to be the case. For example, autotetraploid *Secale cereale* has a mean DNA content per chromosome of 0.64 pg and a mean prophase I quadrivalent

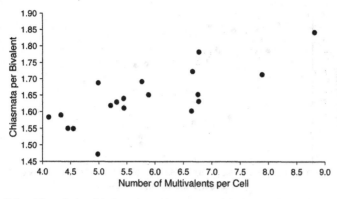

Figure 8.1. The relationship between chiasmata per bivalent in diploids and the number of multivalents per cell in their tetraploid derivatives in *Hyocyamus muticus* (based on Srivastava et al., 1992).

frequency of 0.42 (Chatterjee and Jenkins, 1993), whereas autotetraploid *Triticum monococcum* has a mean DNA content of 0.46 pg and a mean IV frequency of 0.80 (Gillies et al., 1987), and autotetraploid *Crepis capillaris* has a mean DNA content of 0.39 pg per chromosome and mean IV frequency of 0.86 (Jones and Vincent, 1994).

Fertility reduction

The formation of multivalents by autotetraploids typically results in abnormal chromosome segregation and reduced fertility relative to their diploid prototypes. The relationship between multivalent formation and pollen fertility is well illustrated in 19 genotypes of *Hyoscyamus muticus* (Srivastava et al., 1992). There is a statistically significant negative correlation ($r = 0.47$) between these variables (figure 8.2). In this species, there also is a significant negative relationship between chiasma frequencies in diploids and fertility in autopolyploids ($r = -0.76$) This was to be expected because, as noted above, diploids with relatively low chiasma frequencies produce tetraploids with relatively low multivalent frequencies. A negative correlation between chiasma frequencies in diploids and fertility in autopolyploids also was described in *Avena strigosa* (Zadoo et al., 1989) and in *Arachis hypogaea* (Singh, 1986).

If there is a negative correlation between multivalent formation and fertility, it follows that selection for higher fertility may be accompanied by increased bivalent formation. Bremer and Bremer-Reinders (1954) selected for higher fertility in synthetic tetraploid rye and found an increase in equal anaphase I distributions of chromosomes (14 and 14) from 50% in the third generation to 75% in the sixth generation. The frequency of anaphase I laggards in the sixth generation was only one-third that in the first generation. Hilpert (1957) also succeeded in increasing meiotic regularity in rye (more bivalents and fewer univalents and multivalents) by three generations of selection for higher seed set.

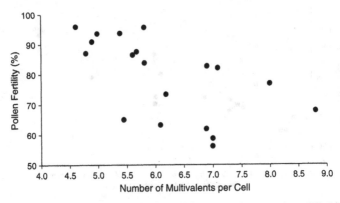

Figure 8.2.   The relationship between pollen fertility and the number of multivalents per cell in tetraploid genotypes of *Hyocyamus muticus* (based on Srivastava et al., 1992).

It is noteworthy that fertility gains in some induced autotetraploids may be achieved without in increase in bivalent formation. For example, Crowley and Rees (1968) reported that after seven generations of selection for higher fertility plants of *Lolium perenne* had a higher frequency of quadrivalents than unselected plants. Similarly, in some lines of rye improved fertility was accompanied by a higher frequency of quadrivalents (Müntzing, 1951; Hazarika and Rees, 1967).

Dudley and Alexander (1969) increased the percentage of seed set (fertility) by 30% through nine generations of selection in a population of synthetic, maize autotetraploids. However, the chromosome pairing relationships observed in the first generation (means of univalents, 0.03/cell; bivalents, 8.29/cell; trivalents, 0.00/cell; quadrivalents, 5.84/cell) essentially were unchanged, whereas fertility increased (Mastenbroek et al., 1982). Presumably, the increase in fertility was due to an alteration in disjunction patterns that yielded more balanced gametes.

GENETIC CONSEQUENCES OF AUTOPOLYPLOIDY

Patterns of segregation

Autotetraploids have a very different system of inheritance (tetrasomic inheritance) than diploids (disomic inheritance), as reviewed by Sybenga (1992). In addition to the two homozygous states, autotetraploids have three possible heterozygous states for a given locus: AAAa (triplex), AAaa (duplex), and Aaaa (simplex). If the four homologous chromosomes pair and assort at random, the expected gametic ratio in a duplex heterozygote will be 1 AA : 4 Aa : 1 aa. This ratio will be obtained only if there is no crossing over between the gene in question and the centromere. This situation produces what is referred to as chromosome segregation.

With chromosome segregation, the crossing of two duplex heterozygotes yields a genotypic ratio of 1 AAAA:8 AAAa:18 Aaaa:8 Aaaa:1 aaaa. This contrasts with a ratio of 1 : 4 : 6 : 4 : 1 ratio for digenic–disomic inheritance or 1 AA : 2 Aa:1 aa for monogenic–disomic inheritance.

When crossing over occurs between the gene and the centromere, chromatid segregation of alleles ensues. The gametic ratio depends on the pattern of crossing over within the quadrivalent and the type of centromere segregation (Marsden et al., 1987). There are three types of segregation: opposite, tangential, and juxtaposed. If crossovers are random and the three segregation patterns occur with equal frequency, the expected gametic ratio in a duplex heterozygote will be 3 AA:8 Aa:3 aa. However, if a single crossover occurs between each pair of chromosomes (i.e., four per quadrivalent), and if the three centromere segregation patterns occur with equal frequency, then the gametic ratio will be 2 AA:5 Aa:2 aa.

The pattern of tetrasomic inheritance is impossible to predict without knowledge of crossover and bivalent and quadrivalent frequencies. To allow more accurate insights into the pattern of tetrasomic inheritance, Jackson and Jackson (1996) formulated a "meiotic configuration" method for estimating gametic frequencies that uses chiasma frequencies inferred from the frequencies of meiotic configurations (viz., bivalents, chain quadrivalents, and ring bivalents). For the most part, this approach gave a better match with the phenotypic ratios at several loci in *Solanum* polyploids than either model of chromosome or chromatid segregation.

Inheritance studies using morphological traits are few owing to a paucity of single-gene polymorphisms (Soltis and Soltis, 1993). Allozymes have become the prime vehicle for studying tetrasomic inheritance, because they offer a rich source of diversity and because it is possible to distinguish among doses of gene products.

Studies on autotetraploid genetics test data for best fit to either chromosome segregation or chromatid segregation. For the most part, the markers chosen behave in accord with chromosome segregation. For example, Shore (1991b) investigated the inheritance of *Adh-2* in *Turnera ulmifolia*. He crossed *ffss* × *ffff* and the reciprocal. The observed ratio of *ffff* to *fffs* to *ffss* was 20:60:16. In *Heuchera micrantha* (Soltis and Soltis, 1989c) and in *H. grossularifolia* (Wolf et al., 1989), all progeny genotype ratios for four isozyme loci were consistent with chromosome segregation. Similarly, in *Aster kantoensis,* all progeny genotype ratios for four isozyme loci were consistent with chromosome segregation (Maki et al., 1996). In *Maclura pomifera,* the majority of genotype ratios across seven isozyme loci fit the expectations for chromosome segregation (Laushman et al., 1996). Two of three isozyme markers in alfalfa showed chromosomal segregation (Quiros, 1982). Other demonstrations of tetrasomic inheritance in polyploids using isozyme markers include *Lotus alpinus* and *L. corniculatus* (Gauthier et al., 1998) and *Centaurea jacea* (Hardy et al., 2000).

Inbreeding depression

After one generation of self-fertilization, full homozygosity (AA or aa) at a heterozygous locus in diploids increases by 50% whereas full homozygosity

(AAAA or aaaa) in tetraploids increases at a much slower rate (17–21%), depending on the pattern of segregation (Haldane, 1930; Wright, 1938b; Parsons, 1959). It will take autotetraploids an average 3.8 generations of selfing to achieve a 50% reduction in heterozygosity versus 1 generation in diploids.

Lande and Schemske (1985) predicted that the level of inbreeding depression due to recessive lethal and sublethal alleles will be less within populations of autotetraploids than diploids. The equilibrium inbreeding depression of a tetraploid should be nearly half that of its diploid progenitor. No difference between ploidal level is predicted for alleles that are partially recessive. If inbreeding depression is due to a loss of allelic interactions rather than to the exposure of deleterious recessive alleles, inbreeding depression may actually be higher in tetraploids (Bever and Felber, 1992).

The most direct test of Lande and Schemske's (1985) prediction in wild plants was made on *Chamerion angustifolium* (Onagraceae). Husband and Schemske (1997) estimated inbreeding depression in the greenhouse for two diploid and three tetraploid populations as judged by seed set and germination, plant survival, and dry mass at 9 weeks (figure 8.3). The tetraploids exhibited less inbreeding depression than the diploids, the cumulative inbreeding depression calculated from these stages averaging 0.95 in diploids versus 0.67 in tetraploids (figure 8.4). The two ploidal levels have similar levels of selfing in nature (close to 0.45), so differences between them cannot be attributed to divergent mating systems.

The relationship between ploidal level and inbreeding depression has been studied in several crops. Synthetic tetraploids of maize (Alexander, 1960), crested wheatgrass (Dewey, 1969), and clover (Townsend and Remmenga, 1968) have less inbreeding depression than their diploid prototypes. Conversely, tetraploids have relatively high levels of inbreeding depression in orchardgrass (Kalton et al., 1952). Regardless of the relationship, polyploidy affects inbreeding depression in all of the aforementioned species.

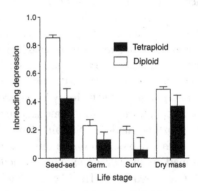

Figure 8.3.   Mean inbreeding depression for diploid and tetraploid populations of *Chamerion angustifolium* at each of four life stages: seed set, germination, survival, and dry mass. Redrawn from Husband and Schemske (1997), with permission of the Society for the Study of Evolution.

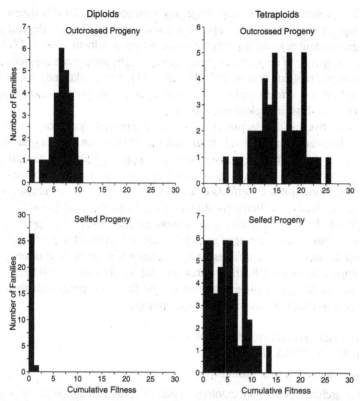

Figure 8.4. Distribution of cumulative fitnesses for selfed and outcrossed progeny from 29 diploid maternal families and 41 tetraploid maternal families in *Chamerion angustifolium*. Redrawn from Husband and Schemske (1997), with permission of the Society for the Study of Evolution.

There also are some basic differences between diploids and autotetraploids with regard to heterosis. Maximum heterosis in diploids is reached after one generation of crossing disparate plants. Conversely, in autotetraploids heterosis is progressive and is not reached until the second generation of crossing or even later, depending on the level of inbreeding in the parents (Bingham et al., 1994).

The behavior of genes in populations

The behavior of genes in autopolyploid populations departs from that in diploid populations in three important ways. First, Hardy-Weinberg proportions in panmictic tetraploid populations are attained only gradually, as opposed to in one generation in diploid populations (Haldane, 1930; Bennett, 1954). Second, heterozygosity decays at a slower rate in autotetraploid populations than in diploid populations with the same gene frequencies (Moody et al., 1993). The variance in autotetraploid gene frequencies (p and q) from one generation to another is $pq/4N_e$, where N_e is the variance effective size (Crow and Kimura,

1970). This contrasts with a gene frequency variance of $pq/2N_e$ for diploids. The change in population phenotype for a given gene frequency is dependent on the gene frequency and the type of genic action at a locus (Rowe, 1986). Third, the response to selection in populations with the same gene frequencies is less in autotetraploids than in diploids (Hill, 1971; Savchenko and Rokitskii, 1975). Indeed, the effectiveness of selection against a recessive or dominant gene is twice as high in diploid populations.

If the effectiveness of selection is higher in diploid populations, does it follow that polyploids adapt more slowly than diploids? It has often been argued that the answer is yes. Stebbins (1950) suggested that polyploids may have great difficulty evolving new adaptive genes combinations because the high level of gene duplication dilutes the effects of new mutations and gene combinations. This issue received little attention until recently, when Otto and Whitton (2000) argued (on the basis of population genetic models) that the rate of adaptive evolution is not necessarily slower in polyploids. Indeed, in small and moderately sized populations, the rate of fitness increase depends more on how often mutations appear and are established within populations than on the efficiency of selection. The rate of adaptive evolution can be faster for polyploids than in diploids if beneficial alleles are partially dominant.

GENETIC AND CHROMOSOMAL PROPERTIES OF NATURAL AUTOPOLYPLOIDS

Chromosome pairing relationships

Many researchers have sought cytological evidence for autotetraploidy in polyploid populations. Approximation of the expected distributions of univalents, bivalents, and quadrivalents based on random chromosome pairing (Sybenga, 1975; Jackson and Casey, 1982; Jackson and Hauber, 1982) would support an inference of autoploidy. Such approximations have been observed in several species, including *Turnera ulmifolia* (Shore, 1991a). The occurrence of tetrasomic inheritance at three isozyme loci (Shore, 1991b) supports the idea that populations of this species are autotetraploids. Similarly, in *Heuchera grossulariifolia* chromosome configurations were consistent with expectations for an autotetraploid, and segregation patterns of four isozyme markers were in accord with tetrasomic inheritance (Wolf et al., 1989).

Cytological evidence for autopolyploidy also comes from comparisons of natural polyploids with synthetic polyploids. In *Turnera ulmifolia* (Shore, 1991a) and *Lathyrus pratensis* (Khawaja et al., 1997), meiotic pairing configurations and fertility in natural autotetraploids are similar to those of synthetic polyploids. This similarity supports an inference of autopolyploidy.

Occasionally tetraploid populations are found that incorporate features of two diploid subspecies and probably are derived by hybridization between them. May these populations be autotetraploid even though they are derived from divergent taxa? The answer is yes, if the chromosomes of the diploid taxa were fully homologous. Hauber (1986) discovered such a situation in *Haplopappus spinulosus.* A series of tetraploid populations in southeastern Colorado were morphologically intermediate between the diploid subspecies *glaberrimus* and *spinulosus.* Chromosome pairing configurations were consistent with expectations for an autotetraploid. As anticipated, synthetic diploid F_1 hybrids exhibited normal meiotic behavior. Tetrasomic inheritance at the *Pgi* locus provided additional evidence of an autoploid nature. Autotetraploid pairing also has been described in induced tetraploids based upon subspecific hybrids of *Dactylis* (McCollum, 1958).

Complete or partial bivalent formation in doubled hybrids will occur if their chromosome complements are differentiated. This is seen in doubled hybrids between *Gossypium* species (Iyengar, 1944) and between *Solanum* species (Swaminathan and Howard, 1953) and between rice varieties (Oka et al., 1954). The assumption in these and other cases is that bivalent formation (preferential pairing) involves members of the same genome, that is, identical chromosomes instead of homoeologous chromosomes. To test this assumption, one must be able to identify chromosomes that belong to each parental genome. A doubled hybrid will have two chromosomes from each parent, and thus a given chromosome can pair with one identical chromosome or two homoeologous chromosomes. If pairing occurs at random, the ratio will be 1:2.

Benavente and Orellana (1991) exploited differences in C-banding patterns for chromosome I to determine pairing patterns in doubled hybrids between varieties and species of *Secale.* They found a tendency for identical chromosomes to pair over homoeologous chromosomes. However, preferential association was found between homologous chromosomes in some hybrids, signifying that genetic factors as well as chromosomal factors influenced pairing behavior.

Whatever the level of abnormal pairing in autopolyploids, the potential exists for an improvement. The most abnormal pairing is expected in newly emergent autopolyploids, whose chromosome behavior should approach that of synthetic polyploids. Subsequent selection for higher fertility may be accompanied by an increased level of bivalent formation, as shown in rye (Bremer and Bremer-Reinders, 1954; Hilpert, 1957).

Genome size

Genome size is a variable that also has the potential to evolve in autotetraploids. Indeed, it is not unusual to find that natural autopolyploids have less DNA per cell than their synthetic counterparts with the same genomic constitution. The DNA contents of colchicine-induced autotetraploids of *Chrysanthemum viscosum* and *C. leucanthemum* were approximately twice that of their diploid prototypes and were at least 10% more than their natural counterparts (Dowrick and El-Bayoumi, 1969). In *Cochlearia pyrenaica* natural autotetraploids have 9.5% less DNA than do induced autotetraploids (Gupta, 1981).

An interesting case of DNA loss in colchitetraploids was reported in *Phlox drummondii.* Raina et al. (1994) found that the initial generation of autotetraploids had 17% less DNA than expected (twice the value of its diploid progenitor). Additional DNA was lost in the next two generations bring the overall reduction to 25%. Moreover, the loss was achieved by an equal decrement in all chromosomes within the complement. The reduction in genome size was accompanied by an increase in seed set from 30% to 66%, which Raina et al. (1994) interpreted as evidence that the lose of DNA was adaptive.

There are several examples of allotetraploids with less than the expected genome size, as discussed in chapter 9. Thus, the aforementioned autotetraploids are not unusual in this respect.

Genetic variation

Comparisons of genetic variation and heterozygosity in diploids and their autotetraploid derivatives have been made in a few species using allozymes. We would expect autotetraploids to have higher levels of heterozygosity, all else being equal, because a given individual has a greater chance of carrying different alleles (due to tetrasomic inheritance) than does a diploid individual. Indeed, a given tetraploid can have up to four different alleles. Unusually high levels of heterozygosity were observed in *Tolmiea menziesii*, where 39% of 678 plants exhibited three or four alleles at least one of the eight loci considered (Soltis and Soltis, 1989a).

Three variables have been used to compare genetic variation in the two ploidal levels: proportion of polymorphic loci, mean number of alleles per polymorphic locus, and observed heterozygosity. The values for five species are presented in table 8.3. In general, the autotetraploids are more variable than their diploid counterparts. Only *Turnera ulmifolia* var. *intermedia* contradicts the expected pattern. However, the populations studied were restricted to islands and thus may reflect the results of a genetic bottleneck during colonization and perhaps subsequently.

Ploidal levels may differ substantially in their levels of genetic variation. For

Table 8.3. Allozyme variation within diploid and autotetraploid populations

Species	A		P		H		Reference
	2×	4×	2×	4×	2×	4×	
Tolmiea menziesii	3.0	3.75	0.07	0.24	0.07	0.24	Soltis and Soltis (1989c)
Heuchera micrantha	1.41	1.64	0.24	0.38	0.07	0.15	Ness et al. (1989)
Heuchera grossularifolia	1.35	1.55	0.24	0.31	0.06	0.1	Wolf et al. (1990)
Dactylis glomerata	1.51	2.36	0.70	0.80	0.17	0.43	Lumaret (1985, 1988)
Turnera ulmifolia	2.20	2.00	0.46	0.40	0.11	0.07	Shore (1991b)

A, mean number of alleles per locus; H, mean observed heterozygosity; P, proportions of polymorphic loci.

example, the proportion of polymorphic loci is 0.24 in diploid *Tolmiea menziesii* versus 0.41 in the autotetraploid (Soltis and Soltis, 1989a). Observed heterozygosity in this species is 0.07 in the diploid versus 0.24 in the autotetraploid. In *Dactylis glomerata* the mean number of alleles per polymorphic locus is 1.51 in diploids versus 2.36 in autotetraploids (Lumaret, 1985, 1988).

OVERVIEW

Autopolyploids have unique meiotic and genetic signatures that differentiate them from allopolyploids. Because autopolyploids have more than two homologues capable of pairing, multivalents are common during the first meiotic division. There is a tendency for homologues with higher chiasma frequencies to form multivalents at a higher rate than homologues with low frequencies. Larger chromosomes tend to have more chiasmata than smaller ones, so within a chromosome complement larger chromosomes are more likely to be involved in multivalents. Chiasma formation is under genetic control; and thus polyploid lines derived from diploids with relatively high levels of chiasma formation tend to have more multivalents than lines derived from diploids with relatively few chiasmata.

The formation of multivalents typically leads to abnormal chromosome segregation and reduced fertility. Accordingly, genotypes with relatively few chiasmata tend to have higher fertilities than genotypes with relatively many chaismata. Selection for increased fertility has, in some species, resulted in an increase in bivalent formation. Surprisingly, in other species fertility gains were achieved without an increase in bivalent formation.

When two heterozygotes are crossed, autotetraploids deviate from the 1 AA:2 Aa:1 aa genotypic ratio for monogenic-disomic inheritance that characterizes their diploid progenitor. The ratio will be 1 AA:8 AAAa:18 AAaa: 8Aaaa:1 aa if the parental genotypes are AAaa and if there is no crossing over between the gene in question and the centromere. With crossing over other ratios occur that depend on the crossover frequency. Genotypic ratios consistent with tetrasomic inheritance have been described in autotetraploids.

The level of inbreeding depression due to recessive lethal and sublethal genes is expected to be less within populations of autopolyploids than diploids. This is because after one generation of self-fertilization full homozygosity (AA or aa) at a heterozygous locus increases by 50% in diploids, whereas full homozygosity in autopolyploids increases at a much slower rate. In autotetraploids it would take an average of 3.8 generations of selfing to achieve a 50% reduction in heterozygosity versus one generation in a heterozygous diploid. Empirical studies involving self-fertilization show that diploids typically experience higher levels of inbreeding depression than autopolyploids.

The behavior of genes in autopolyploid populations departs from that in diploid populations in three important ways. Heterozygosity decays at a slower rate in autopolyploid populations. Whereas Hardy-Weinberg proportions are attained in one generation of random mating in diploid populations, these proportions are attained only gradually in autopolyploid populations. Finally, the response to selection in populations with the same gene frequencies is less in autopolyploids than in diploids. This difference makes one wonder whether autopolyploids adapt more slowly than their diploid counterparts.

Polyploids are expected to be more heterozygous than diploids at a given allele frequency, because a polyploid individual has more gene doses per locus. Analyses of heterozygosity levels in diploid and autotetraploids typically reveal the expected differential.

There have been many instances when it was not clear whether a taxon is an autopolyploid or an allopolyploid. Chromosome pairing and to a lesser degree genetic ratios have been used to solve this problem, because the multivalents and polysomic inheritance of autopolyploids contrast with the normal pairing and disomic inheritance typically exhibited by allopolyploids. Confirmations of autopolyploidy also have come from comparisions of synthetic autopolyploids with natural polyploids. Surprisingly, some synthetic polyploids have more DNA than their wild counterparts. This suggests that the latter have lost DNA, which may be a form of adaptive evolution.

9

The Evolution of Polyploid Lineages

AUTOPOLYPLOIDY AS A FACTOR IN EVOLUTION

The traditional view has been that autopolyploidy is unimportant because it is maladaptive. Levan (1945) argued that chromosome doubling radically alters the balance of multiple factors acting on a single trait and that this balance may be restored only through hybridization and selection. Stebbins (1971) stated, "Clearly chromosome doubling by itself is not a help but a hindrance to the evolutionary success of higher plants" (p. 126). Thus autopolyploidy was thought to be very rare in natural populations. Indeed, Stebbins (1947, 1950) considered only *Galax aphylla* (now referred to as *G. urceolata*), and Clausen, Keck, and Hiesey (1945) considered only *Biscutella laevigata* and *Zea perennis* in addition to *G. aphylla* to be clear examples of autopolyploids.

The number of races and species thought to be autopolyploid has greatly expanded during the latter half of the twentieth century. Autopolyploidy seems to be rampant in some genera. For example, autopolyploidy occurs in 9 of the 11 species of *Artemisia* subgenus *Tridentatae* (McArthur and Sanderson, 1999) and in species in other subgenera, as well (Stahevitch and Wojtas, 1988). Multiple ploidal levels occur within some *Artemisia* species. For example, diploid, tetraploid, and hexaploid taxa occur in *A. rothrockii;* and diploid, tetraploid, and octaploid taxa occur in *A. cana.*

Multiple ploidal levels also occur in many other species. *Asphodelus ramosus, A. albus,* and *A. macrocarpus* (Diaz Lifante, 1996), *Ambrosia dumosa* (Raven et al., 1968), and *Dalea formosa* (Spellenberg, 1981) have $2x$, $4x$, and $6x$ races. *Eragrostis cambessediana* contains $2x$, $4x$, and $8x$ races (Hagerup, 1932). *Claytonia perfoliata* contains $2x$, $4x$, $6x$, $8x$, and $10x$ races (Miller, 1976), as does *Galium anisophyllum* (Ehrendorfer, 1965).

Species with the means for asexual reproduction may have taxa with odd ploidal levels as well as even ones. For example, *Allium grayii* has $2x$, $3x$, $4x$, $5x$, and $6x$ population systems (Kurita and Kuroki, 1964). *Hieracium pilosella* ranges from $2x$ to $10x$ with odd ploidal levels in apomictic taxa (Gadella, 1987). Bayer and Stebbins (1987) describe several *Antennaria* species with sexual diploids and apomictic derivatives at higher ploidal levels some of which are odd as in *A. media* ($2x$, $4x$, $7x$, $8x$), *A. fresiana* ($2x$, $4x$, $7x$), and *A. rosea* ($2x$, $3x$, $4x$, $5x$).

Polyploids may have chromosome numbers that deviate from the basic ploidal levels. This aneuploidy is especially pronounced in apomictics. One classic example involves *Bouteloua curtipedula*, where $2n = 20$ (Gould and Kapadia, 1962, 1964; Gould, 1966). Variety *curtipedula* has the following chromosome numbers: 40–46, 48, 50–56, 58–60, 62, and 64. Variety *caespitosa* has the following numbers: 58, 60, 62, 64, 69, 72, 77–80, 82, 84–88, 90–92, 94, 96, 98, 100, 102, 103. The dividing line between sexuality and apomixis lies at approximately $2n = 50$.

Whereas autopolyploids can arise from diploids and autopolyploids with fewer chromosome sets, chromosome doubling also generates new population systems within allopolyploid species. This possibility may have been realized in *Dahlia* (Gatt et al., 1998).

Before the advent of molecular markers, the prime criteria of autopolyploidy were a multiple of the diploid chromosome number, quadrivalent pairing, reduction in fertility, and tetrasomic inheritance. Whereas these criteria are still operative, we would expect that autopolyploids would share most nuclear alleles with the diploid populations and that the shared alleles occur with similar frequencies (Soltis and Soltis, 1993). We also would expect that autopolyploids have the same or very similar chloroplast DNA markers as their diploid counterparts. We must recognize, however, that as autopolyploids age, their nuclear and chloroplast genomes will diverge from those of their progenitors and thus not quite fit the aforementioned expectations.

Recent studies have uncovered autopolyploid population systems in several species. One such species is *Heuchera grossularifolia*. This species is composed of diploid ($2n = 14$) and tetraploid populations ($2n = 28$). Wolf, Soltis, and Soltis (1990) analyzed variation at 12 isozyme loci and found a total of 44 alleles. Thirty-four of the alleles were shared by both cytotypes; three were unique to the diploids, and seven were unique to the tetraploids (table 9.1).

Wolf, Soltis, and Soltis (1990) also analyzed a total of 170 cpDNA restriction sites. In general, the fragment patterns of the cytotypes were quite similar. An important finding was that variation present among diploid populations was repeated among tetraploid populations. The phylogenetic tree generated from restriction site mutations indicates that autotetraploidy occurred independently at least three times.

Additional evidence of autopolyploidy in *Heuchera grossularifolia* comes from genetic and cytogenetic studies. The tetraploids exhibit multivalent fre-

Table 9.1.   Distribution of alleles among diploid and tetraploid
populations of *Heuchera grossularifolia*

Locus	Diploid only	Tetraploid only	Both cytotypes
Pgm-2	2	1	6
Mdh-1	0	1	2
Mdh-2	0	0	1
Mdh-3	0	0	2
Tpi	0	2	3
Idh	0	0	3
Pgi	0	1	6
G3pdh	1	0	2
Cat	0	1	1
Lap	0	0	5
Ald	0	1	2
Sod	0	0	1

Source: Adapted from Wolf et al. (1990).

quencies as expected in an autoploid and also tetrasomic inheritance at isozyme loci (Wolf et al., 1989).

Autopolyploidy also has been well documented in *Tolmiea menziesii*. The cytotypes have very similar allozyme frequencies, although tetraploid populations have alleles at some loci not present in diploid populations (Soltis and Soltis, 1989a). The cytotypes have 5S and 18S–25S ribosomal RNA genes of identical repeat length and restriction profile (Soltis and Doyle, 1987). Moreover, diploid and tetraploid populations have identical foliar flavonoids, floral anthocyanins, karyotypes, and chromosomal banding patterns (Soltis and Soltis, 1989a).

The documentation of autopolyploidy has been accompanied by demonstrations of their multiple, independent origins. In addition to *Heuchera grossularifolia,* evidence for at least three independent origins have been presented for *H. micrantha* (Wolf et al., 1990) and for *Plantago media* (Van Dijk and Bakx-Schotman, 1997). This evidence is in the form of genetic differentiation among tetraploid populations that corresponds geographically to genetic differentiation among diploid populations.

As autopolyploids age, they are expected to become more divergent from their diploid progenitors and warrant recognition as separate species. We see this in *Solidago,* where the woodland diploid *S. flexilis* apparently has given rise to the sandstone endemic *S. albopilosa* (Esselman and Crawford, 1997). Although the species are morphologically similar, they can be readily distinguished. The species share many alleles, but the tetraploid has some unique ones as well.

Another case in point occurs in *Vaccineum,* where *V. corymbosum* appears to be an autotetraploid derivative of *V. darrowi* (Qu et al., 1998). Although the species differ manifestly in their morphologies and geographical distributions, their affinity is evident in tetraploid hybrids derived from unreduced *V. darrowi* gametes. These hybrids are highly fertile and have tetrasomic inheritance.

ALLOPOLYPLOIDY AS A FACTOR IN EVOLUTION

Polyploidy typically is associated with hybridization, as most polyploid taxa have two distinctive genomes (Grant, 1981). The importance of allopolyploidy as an evolutionary factor traditionally has been attributed to the positive inter-actions, recombination, and/or additive effects of different genomes and to the modification of these genomes by selection.

There are two focal issues related to the study of allopolyploidy. The first involves their ancestry (or genomic constitution). The second involves the genetic and chromosomal fluxes that occur in their early history.

The ancestry of allopolyploids

Prior to 1960 attempts to understand the progenitors of allopolyploids were based on morphological, cytological, and ecogeographical criteria. Because polyploids were expected to be intermediate to their progenitors, diploid species that resembled polyploids or flanked polyploids in many attributes were the usual suspects, especially if they were sympatric with the polyploids. In some instances, interpretations of parentage based on these considerations were corroborated by resynthesizing the wild polyploid through crosses between the putative diploid progenitors. One notable reconstruction was that of *Galeopsis tetrahit* ($2n = 32$) from its putative progenitors, *G. pubescens* ($2n = 16$), and *G. speciosa* (Müntzing, 1930a,b). The new tetraploid not only resembled its wild counterpart; the two were also interfertile. The same can be said of reconstructed *Gilia transmontana* (from *G. minor* and *G. clokeyi*) and *G. malior* (from *G. minor* and *G. aliquanta;* Day, 1965). When each synthetic polyploid was crossed with its natural counterpart, complete chromosome pairing was observed in over 90% of the pollen mother cells, and pollen fertility exceeded 60%.

One notable polyploid complex studied by traditional and modern techniques is the Appalachian *Asplenium* complex. It is composed of three diploid ($n = 36$) and three tetraploid ($n = 72$) species. Wagner (1954) proposed that the tetraploid *A. ebonoides* was derived from the diploids *A. platyneuron* and *A. rhizophyllum,* that the tetraploid *A. bradleyi* was derived from the diploids *A. montanum* and *A. platyneuron,* and that the tetraploid *A. pinnatifidum* was derived from the diploids *A. montanum* and *A. rhizophyllum* (figure 9.1).

Wagner's (1954) presumed parentages were corroborated by comparing the phenolic profiles of diploids and tetraploids (Smith and Levin, 1963; Harborne et al., 1973). *Asplenium rhizophyllum* had four diagnostic compounds *A. montanum* seven compounds, and *A. platyneuron* five compounds. The tetraploids shared the markers of their putative parents. Wagner's views also were corroborated by allozyme studies. Werth et al. (1985) found that each diploid species has a unique allele at three loci, in addition to being rather divergent at poly-

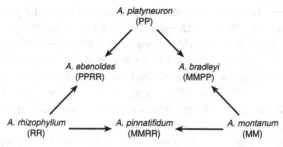

Figure 9.1. Genomic relationships in the Appalachian *Asplenium* complex. Adapted from Wagner (1954).

morphic loci. The tetraploids are heterozygous for the unique alleles of their putative progenitors.

In the 1980s and 1990s, chloroplast DNA, ribosomal DNA and randomly amplified polymorphic DNA (RAPD) markers complemented allozyme markers in diagnoses of polyploid genome constitutions. Moreover, by comparing the cpDNA profiles of polyploids with those of their putative parental species, the maternal parent may be inferred, if the mode of chloroplast inheritance is known. Typically it is maternal in angiosperms.

The genus *Tragopogon* in eastern Washington and western Idaho provides a classic example of how different forms of analyses led to the identification of the genomic constitution of polyploids. On the basis of morphological and cytological considerations, Ownbey (1950) and Ownbey and McCollum (1953) proposed that the tetraploid *Tragopogon miscellus* ($n = 12$) contained the genomes of *T. pratensis* ($n = 6$) and *T. dubius* ($n = 6$) and that the tetraploid *T. mirus* ($n = 12$) contained the genomes of *T. dubius* and *T. porrifolius* ($n = 6$). They also suggested, on the basis of interpopulation variation patterns within each tetraploid, that each tetraploid had arisen more than once.

During the 1970s through 1990s, the ancestries of the two tetraploids have been analyzed with nearly every procedure that has become available. The tetraploids have the diagnostic allozyme alleles of their putative parents (Roose and Gottlieb, 1976; Soltis et al., 1995). They combine the RAPD markers of their putative parents (Cook et al., 1998). Restriction fragment length polymorphism (RFLP) analysis of cpDNA indicated that in the area of Pullman, Washington, *T. miscellus* had *T. dubius* as the maternal parent, whereas elsewhere it had *T. pratensis* as the maternal parent; *T. porrifolius* consistently was the maternal parent of *T. mirus* (Soltis and Soltis, 1989b).

Another informative example of a polyploid complex based on three diploid species is in the genus *Brassica*. The diploids are *B. rapa* (A genome, $n = 10$) *B. nigra* (B genome, $n = 8$), and *B. oleracea* (C genome, $n = 9$). On the basis of hybridization and cytogenetic data, U (1935) proposed that the tetraploid *B. carinata* ($n = 17$) contained the genomes of *B. nigra* and *B. oleracea*, the tetraploid *B. juncea* ($n = 18$) contained the genomes of *B. nigra* and *B. rapa*, and the tetraploid *B. napus* contained the genomes of *B. oleracea* and *B. rapa*.

U's hypothesis has been supported by flavonoid (Dass and Nybom, 1967), allozyme studies (Coulthart and Densford, 1982; Chen et al., 1989), and nuclear RFLP analyses (Delseny et al., 1990; Song et al., 1988). Restriction site analysis of cpDNA and mtDNA revealed that whereas *B. carinatus* and *B. juncea* each had one maternal parent, *B. napus* had more than two (figure 9.2; Erikson et al., 1983; Song and Osborn, 1992). Some strains had the organelle markers of *B. rapa* or *B. oleracea*, whereas others had chloroplast patterns identical to those of the closely related *B. montana*. However, the latter *B. napus* strains shared a unique mitochondrial pattern. Song and Osborn (1992) proposed that *B. montana* or a close relative might be the ancestor of *B. rapa* and *B. oleracea*. Both nuclear and organelle RFLPs indicate that *B. napus* evolved more than once. The *Brassica* polyploids are second order polyploids, because the diploids themselves are polyploids, as discussed above (Lagercrantz and Lydiate, 1996).

Two approaches have been quite helpful in assessing the relationships between the diploid brassicas. One approach was to make interspecific hybrids that contained one dose of each genome present in each tetraploid and assess the pairing relationships of the disparate chromosome sets. Chromosome pairing in hybrids with AB and BC genomes was poor with a maximum number of bivalents of three or four (Attia and Robbelen, 1986a,b). Conversely, the bivalent number in AC hybrids usually was about seven. The mean chiasma frequency in AC hybrids was about 12 per pollen mother cell versus 5.8 in the AB hybrids and 2.0 in BC hybrids. The relative divergence of the B complement also is evident in the sizes of *B. nigra* chromosomes, which are smaller than those of *B. oleracea* (C genome) and *B. rapa* (A genome).

Relationships among the diploid brassicas also were assessed by comparing their patterns of RFLPs for nuclear DNA. Song et al. (1988) found that two evolutionary pathways apparently were operative in the evolution of the diploid brassicas. *B. nigra* came from one pathway apparently originating in the genus *Sinapis*, whereas *B. oleracea* and *B. rapa* came from the other within *Brassica*. These findings support an earlier study on the restriction patterns of cpDNA in

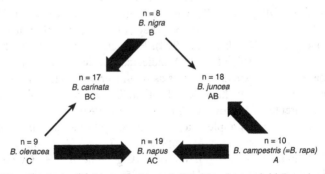

Figure 9.2. Genomic affinities of cultivated diploid and tetraploid *Brassica* species. The maternal parents of the tetraploids are indicated by the thick arrows. The origin of *B. napus* is more involved than shown here (based on Erickson et al., 1983; Palmer et al., 1983; Song and Osborn, 1992).

which Palmer et al. (1983) found that the pattern of *B. nigra* is distinct from the similar patterns of *B. rapa* and *B. oleracea.*

Another well-studied pair of tetraploids is *Gossypium hirsutum* and *G. barbadense* ($n = 26$). These species share the genomic constitution AADD. Chromosome pairing relationships in interspecific hybrids suggests that the A genome resembles that of *G. herbaceum* ($n = 13$) and that the D genomes resembles that of *G. raimondii* ($n = 13$; Endrizzi et al., 1984). Studies of the chloroplast (Olmstead and Palmer, 1991) and mitochondrial genomes (Bland et al., 1985) indicate that the maternal genome of *N. tabacum* was donated by a species close to *N. sylvestris.* The polyploids have nearly additive genome sizes regarding their diploid ancestors, which themselves differ by almost a factor of two. The 2C DNA values for the A, D, and AD genomes are 3.8, 2, and 5.8 pg, respectively (Bennett et al., 1982). The diploids have similar amounts of single-copy DNA, indicating that genome size differences were achieved through the accumulation and elimination of repetitive sequences (Geever et al., 1989). The putative A × D polyploidization event occurred subsequent to the transoceanic migration of the maternal A genome ancestor to the New World one to two million years ago (Wendel, 1989; Wendel and Albert, 1992).

The diagnosis of the genomic constitutions of polyploids recently has been facilitated by the use of genomic *in situ* hybridization (GISH). This technique allows the chromosomes from different parents or progenitors to be "painted" different colors. Whether parental genomes have alternate colors is a function of their evolutionary divergence. Accordingly, it may not be possible to distinguish between genomes of closely related species.

The genomic origin of the amphidiploid *Nicotiana tabacum* ($2n = 48$) has been studied with GISH. The species is a putative derivative of hybridization between *N. sylvestris* (genome S) and a species in the section *Tomentosae* of the subgenus *Tabacum* (genome T). With GISH, 24 of the chromosomes of *N. tabacum* fluoresced yellow when probed with genomic DNA from *N. sylvestris,* whereas the 24 chromosomes of the T genome fluoresced red (Kenton et al., 1993). When the chromosome complement of *N. tabacum* was probed with DNA from species with the T genome (*N. tomentosa, N. tomentosiformis,* and *N. otophora*), the strongest signal was given by *N. otophora.* Thus, the genomes in tobacco are most similar to *N. sylvestris* and *N. otophora.*

The phylogeny of the hexaploid *Festuca arundinacea* ($2n = 42$) also was determined using GISH. Humphreys et al. (1995) concluded that two of the three genomes came from the tetraploid *F. arundinacea* var. *glaucescens* and one came from *F. pratensis.* In *Festuca* and *Nicotiana,* the results were in general accord with the proposed phylogenies based on other criteria.

Three other cases where GISH made a significant contribution are worth noting. Taketa et al. (1999) demonstrated that the genome of the diploid *Hordeum marinum* was present in the tetraploids *H. capense* and *H. secalinum* and the hexaploid *H. brachyantherum.* Kenton et al. (1993) reported that the genome of the diploid *Milium vernale* (Poaceae) was present in the tetraploid *M. montianum.* Raina and Mukai (1999) found that the chromosome complement of the tetraploid cultivated peanut (*Arachis hypogaea*) was very similar to that of the tetraploid *A. monticola* and that both tetraploids contained the chromosome complements of the diploids *A. villosa* and *A. ipaensis.*

The availability of several techniques to identify genomes of polyploids does not guarantee that the genomes indeed will be identified. The diploids carrying a given genome may be extinct or undiscovered, or a genome in a polyploid may be a composite of two or more diploid genomes.

One well-studied polyploid whose ancestry remains somewhat unclear is breadwheat *Triticum aestivum.* This species is an allohexaploid ($2n = 42$) with the genomic designation AABBDD. In 1965, there was general consensus that the A genome was derived from a diploid *Triticum,* the B genome from *Aegilops speltoides,* and the D genome from *Ae. squarrosa* (Riley, 1965). The most recent evidence indicates that *T. urartu* is the source of the A genome and confirms that *Ae. squarrosa* is the source of the D genome (Breiman and Grauer, 1995). It also indicates that the mitochondria and chloroplasts of *T. aestivum* and one progenitor of *T. turgidum* (AABB) are derived from the diploid with the B genome. This appears to be *Ae. speltoides* or a very close relative (Wang et al., 1997). Indeed, the latter appears to be the maternal parent of all polyploid wheat. Blake et al. (1999) cloned and sequenced two single-copy DNA sequences from each of the seven chromosomes of the wheat B genome and the homologous sequences from five diploid species in the section *Sitopsis*. Phylogenetic comparisons of sequence data suggest that the wheat B genome diverged from a species in this section subsequent to a major genetic bottleneck.

As discussed above, interspecific differences in chloroplast genomes of diploid species allow one to identify the species that contributed the maternal genome to polyploids. In addition to their role in documenting the maternal genome phylogeny, chloroplast genome differences also may shed light on the relative age of polyploids sharing the same chloroplast genome. The rationale is that the chloroplast genomes of recently derived polyploids will have diverged less from its diploid prototype than those of older polyploids, because molecular differences accumulate over time. Thus, the genetic distances between recent polyploids and their maternal parent are expected to be less than the distance between older polyploids and their maternal parent. The same rationale holds for mitochondrial DNA.

Genetic distances based jointly on the chloroplast and mitochondrial genomes of different species are known for *Triticum* and its close ally *Aegilops*. To obtain these distances, Wang et. al (1997) used polymerase chain reaction single-strand conformational polymorphism analysis on a 14.0-kb region of cpDNA and a 13.7-kb mtDNA region that contained 10 and 9 structural genes, respectively. One of the more interesting contrasts involves *Ae. squarrosa,* which is the maternal parent of the tetraploids *Ae. cylindrica, Ae. crassa,* and *Ae. ventricosa.* The oldest species apparently is *Ae. crassa* (genetic distance of 0.050 relative to *Ae. squarrosa*) and the youngest is *Ae. ventricosa* (genetic distance of 0.006), with *Ae. cylindrica* being of intermediate vintage (genetic distance of 0.015). *Aegilops speltoides* is the maternal parent of Timopheevi and Emmer wheats. The mean distance between Timopheevi and *Ae. speltoides* is 0.012, whereas that between Emmer and *Ae. speltoides* is 0.067, indicating that Timopheevi is of more recent origin.

The recurrent formation of allopolyploids

If two diploid species are in contact in several localities and are capable of spawning hybrids that in turn produce unreduced gametes, the potential exists for the recurrent origin of polyploids. Below I first consider factors affecting the likelihood of these events and then consider examples of multiple origins.

THEORETICAL CONSIDERATIONS.   The incidence of multiple origins of a polyploid is expected to be positively correlated with the magnitude of parental species sympatry and the levels of overlap in their ecological tolerances and reproductive phenologies. The incidence also is likely to be positively correlated with the levels of progenitor cross-pollination and cross-compatibility, the vigor of interspecific hybrids, and the penchant for these hybrids to generate unreduced gametes.

The number of geographically independent origins that a species actually has is likely to be well below the number of hybridization and chromosome doubling episodes in different populations. One reason is that the number of polyploids at a given site may be insufficient for reproduction. At the extreme, a single outcrossing polyploid would have no mate. Even if there were a few such plants, their reproductive success is likely to be low unless they were near each other, and even then, the prospects are not good unless there was some mechanism to promote assortative mating.

A second reason why the number of independent origins is likely to be well below the number of hybridization and chromosome doubling episodes in different populations is the extinction of newly established polyploid populations. There is no reason to assume that every new population will expand and become an integral part of the species. Local extinction may be much higher in raw polyploids than in long-established polyploids, because various genetic adjustments may be a prerequisite for success, as discussed below.

How do we know if a polyploid originated more than once? It is tempting to conclude that if a polyploid has more than four alleles at a nuclear locus and more than one chloroplast haplotype, it has arisen more than once. The logic is as follows: if two diploid parents are heterozygous for different alleles at a given locus, the maximum number of alleles that a tetraploid plant could contain is four.

The problem with this reasoning is that it assumes that a polyploid species is derived from one individual. This is not likely to be the case, and in obligate outbreeders it could not be the case. Rather, it is likely that at a given locale, more than one polyploid is formed within a generation, most episodes involving different parentages. Thus, a local population of polyploids could at the outset receive the nuclear and chloroplast genomes of several members of each diploid species. If these genomes differed among conspecific diploids, a local population could have more than four alleles at homoeologous loci and more than one chloroplast haplotype. More than one chloroplast haplotype also would accrue if some polyploids carried the chloroplast genome of one parent and others the chloroplast genome of the other parent.

The best evidence for multiple independent origins is a concordance between geographical variation patterns of the diploids and polyploids. Alleles or haplotypes present in the diploids in only one area would occur in the polyploids of that area, and alleles present in the diploids in another area would occur in the polyploids of that area. Multilocus genotypes also may display such correspondence.

In practice, the notion of recurrent origin often is embraced when there is substantial divergence between populations of polyploids. The case becomes more convincing as the classes of markers showing interpopulation discontinuities increases.

As our repertoire of genetic markers expands, it is becoming increasingly apparent that the recurrent origin of allopolyploids may be the rule rather than the exception (Soltis et al. 1995). Most reports of recurrent origins probably underestimate the actual number of recurrences, because they are based on incomplete sampling throughout the areas of sympatry between the polyploids and their progenitors and because they are dependent on a relatively small number of markers.

EXAMPLES OF RECURRENT EVOLUTION.   The notion that polyploids arise through a single episode of hybridization and chromosome doubling was first challenged by Ownbey and McCollum (1953) regarding the tetraploids *Tragopogon miscellus* and *T. mirus*. On the basis of interpopulation variation patterns, they suggested that each tetraploid had arisen more than once. Subsequent studies have proven them correct. Interpopulation differences in the parental taxa are in geographical congruence with differences in the polyploid populations for nuclear and cytoplasmic markers. Soltis and Soltis (1999) and associates have shown that *T. miscellus* has arisen as many as 20 times, and *T. mirus* as many as 12 times. Indeed, multiple origins have even occurred within single small towns in less than 70 years (Soltis et al., 1995; Cook et al., 1998).

The diploid relatives of many polyploid species with large circumboreal or circumpolar distributions often have large distributions (Ehrendorfer, 1980). *Draba* is one genus with such distributions where the latest battery of genetic markers have been brought to bear on the origin of polyploids. Patterns of electrophoretic variation in the hexaploid *D. norvegica* indicate a minimum of three and probably 13 independent origins (Brochmann et al., 1992a). The hexaploid *D. lactea* has a dual origin, with one of its genomes coming from one or the other of two intersterile races of the diploid *D. fladnizensis*, as judged from rDNA restriction profiles (Brochmann et al., 1992b). Multiple origins also have been established for the 16-ploid *D. corymbosa* (Brochmann et al., 1992a) and the octoploid *D. cacuminum* (Brochmann et al., 1992b) on local geographical scales. The latter has evolved at least three times, apparently through hybridization between *D. norvegica* and *D. fladnizensis*. Studies of the aforementioned *Draba* polyploids over a wider area are likely to show evidence of additional recurrences, because the populations studied came from a small geographical area and because the sample sizes were small.

Recurrent origins also have been demonstrated in the segmental allotetraploid *Glycine tabacina* ($2n = 80$). There are two types of soybeans that differ in restriction endonuclease maps of the 18S–26S nuclear rDNA locus, chloroplast DNA haplotype, and morphology (Singh et al., 1987; Hymowitz et al., 1998). One of the two *G. tabacina* taxa (BBB_2B_2) contains eight different chloroplast haplotypes, six of which are identical in their restriction maps to haplotypes present in diploid species of the B-genome assemblage (Doyle et al., 1990). This suggests that *G. tabacina* originated at least six different times. Doyle et al. (1999) analyzed DNA sequence variation at one homoeologous histone H3-D locus, and identified three alleles each also found in different diploid species of the B-genome assemblage. This information further supports the contention of multiple polyploid origins, in this case at least three, assuming no progenitor was heterozygous at this locus.

Glycine tabacina is notable because of the extensive gene exchanges that apparently have occurred among its lineages (Doyle et al., 1999). This is suggested by the lack of concordance between the geographical variation patterns of chloroplast haplotypes, histone alleles, and two isozyme loci. A large number of multilocus genotypes occur within a morphologically unified but widely distributed species that is predominantly self-fertilizing.

Polyploids within *G. tomentella* also appear to be the products of recurrent origins. The species contains tetraploid ($2n = 80$) and aneutetraploid ($2n = 78$) populations. Based on protein and restriction fragment patterns, cytogenetic observations, and the properties of synthetic tetraploids, Kollipara et al. (1994) identified three distinct aneutetraploid groups and four distinct tetraploid groups that are based in large measure on different genomic constitutions.

THE TRANSFIGURATION OF ALLOPOLYPLOID GENOMES

Conventional wisdom held that allopolyploids contained distinct genomes that confer elevated levels of heterozygosity but that remain independent entities (Stebbins, 1971). Recent studies have demonstrated that this is not necessarily the case. Indeed, polyploids may undergo intergenomic chromosomal reorganization, concerted evolution, and gene silencing. In some polyploids (e.g., maize, *Brassica*), reorganization and gene silencing may be so extensive that the genome is structured as a diploid.

Chromosomal rearrangements

GISH showed that extensive intergenomic chromosome reorganization has occurred in the allotetraploid *Nicotiana tabacum* (Kenton et al., 1993). Up to nine homozygous translocations have occurred between the S and T genomes (figure 9.3). Five intergenomic translocations were found in allotetraploid *Avena*

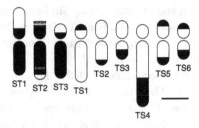

Figure 9.3.   Diagrammatic portrayals of nine intergenomic chromosome translocations in tobacco. The black and white areas represent the S and T genomes, respectively. Redrawn from Kenton et al. (1993), with permission of Springer-Verlag.

maroccana, and eight were found in *A. sativa* (Jellen et al., 1994). RFLP analysis in soybean, which is a tetraploid, also has revealed many translocations (Shoemaker et al., 1996).

Two different types of intergenomic translocations are recognized by Jiang and Gill (1994). On the one hand, there are species-specific translocations that are found in every population of a polyploid. On the other hand, there are random translocations that involve different chromosomes in different populations. Species-specific translocations may reflect reorganizations that must occur to restore imbalances created by the union of a male nuclear genome and female nuclear and cytoplasmic genomes in a newly formed polyploid. Jiang and Gill described species-specific translocations in two polyploid wheats. One translocated chromosome in tobacco (ST1 of figure 9.3) may fall into this category.

Chromosomal rearrangements also may occur *within* the genomes of allopolyploids after chromosome doubling. RFLP mapping studies of Brubaker et al. (1999) indicate that the *Gossypium* allotetraploids *G. hirsutum* and *G. barbadense* (AADD) have accumulated a minimum of six inversions and two translocations since the A and D genomes of their progenitors were combined one to two million years ago. Five of the inversions involved the D genome and one the A. The two translocations involve the A genome. The rate of fixation for these rearrangements is estimated at 2.6–5.2 events per million years.

Because the genomes of allopolyploids are more structurally divergent than the same genomes in the diploid progenitors, chromosome pairing in diploid AD hybrids is more complete than is pairing in AD haploids derived from the tetraploids (Mursal and Endrizzi, 1976). Specifically, allotetraploid-derived haploids average less than one bivalent at metaphase I, whereas diploid hybrids average over six bivalents.

Brubaker et al. (1999) also demonstrated that the recombination rates differed between cotton ploidal levels. Specifically, the D genome in the tetraploids had a recombination rate that was 58.5% larger than in its diploid counterpart. The recombination rate in A genome was 51.5% greater than in the diploid. It is noteworthy that even though the A genome is nearly twice as large as the

D genome they had similar recombination rates in both diploid and tetraploid species.

Intergenomic sequence transfer

The union of multiple genomes in polyploids provides an opportunity for intergenomic sequence interaction and evolution. The strongest evidence for this comes from *Gossypium* tetraploids that have two doses of the A genome and two doses of the D genome. Zhao et al. (1995, 1998) showed that of 83 non-cross-hybridizing repetitive DNAs isolated from *G. hirsutum,* about 75% were largely restricted to the A genome and only 5% were largely restricted to the D genome. Most families that were restricted to the A genome at the diploid level were present not only on A genome chromosomes in *G. hirsutum* but also were present on D genome chromosomes as well. This indicates that since polyploid formation, the D genome has been colonized by alien genome- specific sequences.

The story of intergenomic sequence transfer in *Gossypium* tetraploids has another aspect. Instead of evolving independently, some similar repetitive s equences brought together from different genomes have undergone concerted evolution. Concerted evolution refers to the nonindependent evolution of sequences at multiple loci that leads to a greater similarity of repetitive sequences within species (Dover, 1982). *Gossypium* tetraploids received two ribosomal DNA genes from their A genome progenitor and two rDNA genes from their D genome progenitor. Internal transcribed spacer (ITS) sequences between the 5S and 18S rDNAs have become identical to those of the D genome in most AADD tetraploids. However, in *G. mustelinum* these sequences have concerted to an A genome repeat type (Wendel et al., 1995a). Thus, interlocus concerted evolution is bidirectional. In *G. hirsutum,* interspersed repetitive sequences of the A genome have nearly replaced native sequences in the D genome (Hanson et al., 1998).

Interlocus homogenization of alternative rDNA repeat types may be common. This phenomenon has been reported in *Saxifraga* (Brochmann et al., 1996), *Paeonia* (Zhang and Sang, 1999), and *Microseris* (Roelofs et al., 1997) as well as in *Gossypium.*

The colonization of one genome by repetitive elements of another genome in allopolyploids may be driven by transposable elements or by the fact that the repetitive elements themselves are transposable. Matzke and Matzke (1998) contend that polyploids tolerate extensive genome alteration by transposable elements, because polyploid genomes are well buffered due to gene redundancy. Transposition may be promoted by hybridization and chromosome doubling, both of which place genes in a stressful, alien environment. As proposed by Barbara McClintock (1984) in her Nobel Laureate address, one genomic response to shock may be the release of transposable element activity.

Evidence that the interaction of disparate genomes may promote transposable element activity is forthcoming from a study by Liu and Wendel (2000).

They showed that two types of retrotransposons of rice (*copeia*-like and *gypsy*-like) were activated by introgression from *Zizania latifolia.* Activation is expressed as a substantial increase in the copy numbers of the elements. The *copeia*-like element increased 5–10 times in most of the five introgressed lines compared to unintrogressed rice. The presence of less than 1% of the alien genome was sufficient to trigger these transposable elements. It is particularly noteworthy that each of the introgressed lines exhibited heritable phenotypic traits not found in either species (Liu et al., 1999). Thus, it is likely that element activation led to phenotypic novelty. The activity of the elements was ephemeral, lasting only a few generations. Accordingly, the novel traits are expected to persist in the introgressed lines.

Another likely example of hybridization and the mobilization of transposable elements involves *Phlox drummondii.* In areas of hybridization with *P. cuspidata,* one often finds white-flowered plants with red sectors or dots in otherwise red-flowered populations (D.A. Levin, unpublished observations). When these aberrant plants are intercrossed, most progeny are unstable for flower color, but a small fraction are pure white or pure red.

Concerted evolution is not a universal occurrence in polyploids. For example, both parental ITS sequences of nuclear ribosomal DNA are present in allopolyploid peonies that are approximately one million years old (Sang et al., 1995). Similarly, in *Krigia,* both parental sequences of the ITS region have been maintained in an allopolyploid (Kim and Jansen, 1994). Indeed, we do not know how common concerted evolution is in plants of hybrid ancestry.

The silencing of duplicated genes

In addition to molecular interactions mediated by concerted evolutionary processes, duplicated genes may be silenced, diverge in function, or simply be maintained. There are several examples of gene silencing in different venues. For example, in tetraploid *Chenopodium,* null alleles were observed at each of two duplicated leucine aminopeptidase loci (Wilson et al., 1983). The loss of duplicate gene expression at these loci appears to have occurred independently in several different *Chenopodium* polyploids. Silencing also has been observed for one of several genes tested in the hexaploid *Triticum aestivum* (Hart, 1983) and in the fern *Pellaea rufa* (Gastony, 1991).

Contrary to the retention of gene expression in the majority triplicate enzyme-coding loci in hexaploid wheat, there has been a massive diploidization of endosperm proteins (Galali and Feldman, 1984; Feldman et al., 1986). The high-molecular-weight glutinen bands are encoded by the B and D genomes but not the A.

The alteration of duplicate gene expression in polyploids is most widely observed in ribosomal genes that are clustered by the hundreds or thousands of copies at loci known as nucleolus-organizing regions. In many allopolyploids, nucleoli form only form on chromosomes associated with one genome (Pikaard, 2000). In the Triticeae (Dubcovsky and Dvorák, 1995), *Glycine* (Shi et al., 1996), *Brassica* (Snowden et al., 1997), and *Nicotiana* (Volkov et al.,

1999), complete 18S–26S and/or 5S rDNA arrays have been not only silenced after allopolyploidization, but also eliminated. The enforcement of nucleolar dominance apparently is accomplished by selectively silencing one set of RNA genes through the chemical modification of chromatin (Pikaard, 2000). However, why the nucleolus-organizing regions of one genome are active while those of the second are not remains to be determined.

Nucleolar dominance is not a universal feature of amphiploids. ITS repeat types that combine distinct parental motifs in mosaic patterns have been reported in *Gossypium gossypioides* (Wendel et al., 1995b), *Microseris scapigera* (van Houten et al., 1993), and *Microthalaspi perfoliatum* (Mummenhoff et al., 1997).

The most comprehensive study of gene silencing is on the duplicate *PgiC* locus in diploid *Clarkia*. Ford and Gottlieb (1999) showed that *PgiC2* was suppressed independently at least five times subsequent to its evolution. This preceded the radiation of the extant sections of the genus (figure 9.4). The silenced gene is characterized by point mutations, insertions, and deletions (Gottlieb and Ford, 1997). Oddly enough, *PgiC2* is an active gene in the tetraploid *C. gracilis* but not in its diploid ancestors.

Transposable elements may be a causative agent of gene silencing. For example, in hexaploid wheat the loss of glutinen expression at the *Glu-1* locus is due to an 8-kb insertion of a retrotransposon in the coding region (Harberd et al., 1987).

Whereas gene silencing does occur, retention of duplicate gene function seems to be the rule, as noted above and as described in cotton. Cronn et al. (1999) sequenced 16 loci from both genomes of the allotetraploid *Gossypium hirsutum* and from both progenitor (diploid) genomes. They found that most duplicated genes in the allotetraploid evolve independently of each other and at the same rate as those of their diploid progenitors. There has been no increase in deleterious mutations in homoeologous gene pairs.

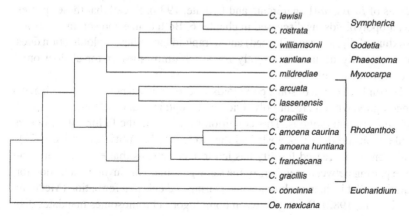

Figure 9.4. Most parsimonious tree for PgiC1 genes of *Clarkia* species with PgiC of *Oenothera mexicana* as the outgroup. Redrawn from Ford and Gottlieb (1999), with permission of the Society for the Study of Evolution.

What fraction of genes is inactivated in allopolyploids? As reviewed by Comai (2000), the work on selected genes suggests that this fraction is quite small. To better address this issue, Comai used an alternative approach that samples more genes, namely, comparative mutagenesis. The ratio of induced mutation rate for the polyploid to the rate for its progenitors can be used to measure gene inactivation. In wheat, the frequency of mutations that reduce or abolish chlorophyll synthesis suggests that 20% of the duplicated genes in the allotetraploid were inactivated. A similar level of inactivation was obtain for the allotetraploid *Arabidopsis suecica*. In maize, about 28% of duplicate genes have been inactivated (Ahn and Tanksley, 1993; Gaut and Doebley, 1997).

The traditional view (Ohno, 1970) that duplicate genes may diverge in function after one copy has been released from its functional constraints has not found much support in polyploid plants, although there is evidence of duplicate gene divergence in maize and wheat (Wendel, 2000). Perhaps the paucity of evidence for duplicate gene divergence is not surprising given that null mutation rates may be orders of magnitude greater than beneficial mutation rates (Walsh, 1995). Nevertheless, as Walsh (1995) notes, the probability of fixing a beneficial mutation that brings about a change in gene function is likely to be much greater than the ratio of advantageous to null mutations, especially when population size is very large. Given strong selection and very large population size, the probability may even be a few to several percent.

Diploidization

In some polyploids gene silencing and chromosomal reorganization are so extensive that the genome is structured as a diploid rather than as a polyploid. Inheritance is disomic and chromosomes form bivalents during meiosis. The greater the progression of diploidization the more difficult it is to discern.

Gene silencing has occurred in maize (Rhoades, 1951; Ahn and Tanksley, 1993; Gaut and Doebley, 1997), *Sorghum bicolor* (Whitkus et al., 1992), and species of *Brassica* (Lagercrantz and Lydiate, 1996). Recall that these species are autopolyploids, as discussed in chapter 6. Both maize and sorghum harbor a mixture of diploidized (silenced) and tetraploid loci. Thus, diploidization does not necessarily occur concurrently for all chromosomes or for all loci on a given chromosome.

The diploidization of some polyploids is accomplished not only through gene silencing. Newly emergent polyploids may contain sets of chromosomes that are partially homologous. This condition may lead to the failure of complete bivalent pairing during meiosis and thus to reduced fertility (Stebbins, 1971). Selection for increased fertility has led to mechanisms that enhance chromosome pairing between members of the same genome. This involves selection for mutations that influence chromosome pairing and chiasma formation (McGuire and Dvorák, 1982). Pairing between homologous chromosomes mandates disomic inheritance.

Synaptonemal complex analysis reveals two major strategies by which bivalent formation is achieved in allopolyploids. The first is the correction (by pachytene) of multivalent synapsis at zygotene coupled with the suppression of crossing over between synapsed regions of homoeologous chromosomes. This strategy occurs in *Triticum aestivum* (Holm and Wang, 1988), *T. Timopheevi* (Martinez et al., 1996), *Lotus corniculatus* (Davies et al., 1990), and synthetic amphiploids between *Lolium perenne* and *L. temulentum* (Jenkins and Jimenez, 1995). The second strategy involves the restriction of synapsis to mostly homologous chromosomes at zygotene and the suppression of crossing over at pachytene. This has been observed in *Festuca arundinacea* and *F. gigantea* (Thomas and Thomas, 1993) and in species of *Aegilops* (Cuñado et al., 1996a,b).

Another cytogenetic strategy to limit multivalent formation has been described in *Chrysanthemum* (Watanabe, 1983). It appears that multivalent formation is prevented by the restriction of pairing initiation to a single site per chromosome.

The regulation of chromosome pairing is best understood in hexaploid wheat (genome constitution AABBDD) and its relatives. The diploidlike chromosome pairing in these species is principally influenced by the *Ph1* locus on the long arm of chromosome 5B (Riley, 1960). The dominant *Ph1* allele suppresses pairing between homoeologous chromosomes in favor of pairing between homologous chromosomes. This gene exerts its effect prior to synapsis, affecting the premeiotic alignment of homologous and homoeologous chromosomes (Feldman, 1993). It is believed that *Ph1* acts through its effect on spindle microtubule dynamics.

Chromosome pairing in wheat polyploids also is controlled by the gene *Ph2* that is located on the short arm of chromosome 3B and by several minor loci (Sears, 1976). The mechanism of *Ph2* pairing control differs from that of *Ph1* (Benavente et al., 1998). We are just beginning to appreciate that chromosome pairing in wheat is controlled by cytoplasmic factors as well as by nuclear ones, as shown in cytoplasmic substitution (alloplasmic) lines (Wang et al., 1999).

The chromosomal diploidization of polyploids is best observed when comparing a wild species with a resynthesized counterpart. One such species is *Senecio cambrensis* ($2n = 60$), which was first recorded about 50 years ago (Rosser, 1955). The species arose via hybridization and chromosome doubling between *S. squalidus* and *S. vulgaris*. Ingram and Noltie (1989) showed that wild plants had 28.84 bivalents versus 26.82 in the resynthesized species. The maximum possible number of bivalents is 30. The mechanism enhancing bivalent formation is unknown.

Changes in genome size

The genome size of an allopolyploid is expected to be the sum of the genome sizes of the ancestral species. This expectation is usually, but not always, realized. Some allopolyploids have significantly less DNA than the sum of their parents. Both conditions are seen in *Nicotiana* (Narayan, 1987). The 2C DNA level in the tetraploid *N. rustica* is 14.57 pg/nucleus, which is roughly the sum of the closely related diploids *N. paniculata* (7.78 pg) and *N. undulata* (6.46 pg). Similarly, the genome size of *N. tabacum* (12.18 pg) is roughly the sum of the closely related *N. sylvestris* (5.74 pg) and *N. otophora* (6.54 pg). However, the genome size of *N. arentsii* (12.18 pg) is well below the expected value of 14.02 pg based on 6.46 pg and 7.56 pg for the closely related *N. undulata* and *N. wigandioides,* respectively.

In *Brassica napus,* we have even more direct evidence of a decline in DNA content, because we can compare synthetic polyploids with natural polyploids. The nuclear volume in the natural form is about 60% less than in the synthetic form (Verma and Rees, 1974).

SALTATIONAL GENOME EVOLUTION

I have shown that allopolyploids are not merely receptacles for parental genomes that are expressed in an additive fashion. Rather, the parental genomes may be transfigured. The orthodox view of polyploid evolution was one of gradual change, where major departures from the original genetic architecture of the polyploid took a considerable length of time (Stebbins, 1971). Recently, we have begun to appreciate that this view may be too conservative. Indeed, new polyploids may evolve quite rapidly.

One of the most notable demonstrations of rapid change was by Song et al. (1995) on *Brassica.* They created reciprocal synthetic allopolyploids between *B. rapa* (genome A) and *B. oleracea* (genome C) and between *B. napa* and *B. nigra* (genome B). Thus, they generated two types of hybrids each in two different cytoplasms. The AACC tetraploids have the same nuclear complement as *B. juncea,* and the BBCC tetraploids have the same complement as *B. napus.*

After colchicine doubling, plants from the F_2 generation through the F_4 generation were self-fertilized to produce four lineages extending to the F_5 generation. RFLP comparisons using nuclear DNA probes in the F_2 and F_5 generations revealed a wide range of alterations in the latter generation. On average 9.6% of the restriction fragments had changed in the *B. rapa* $\times$ *B. nigra* line and 8.2% had changed in the reciprocal line. On average 4.1% of the fragments had changed in the *B. rapa* $\times$ *B. oleracea* line, and 3.7% had changed in the reciprocal line. Changes included gains and losses.

In addition to demonstrating rapid genomic evolution, two important findings emerged from this study. First, the greatest changes occurred in the lines (A $\times$ B and B $\times$ A) that have the most divergent genomes (Song et al., 1988; Lagercrantz and Lydiate, 1996). Second, in these lines the paternal genome showed greater divergence from its diploid parent than did the maternal genome (table 9.2). No significant differences were observed in the A $\times$ C and C $\times$ A lines, although the bias was in favor of greater paternal genome differentiation. The results of this study show once again how cytonuclear interactions may affect the character of polyploids. Parentage does make a difference.

Genome evolution in *Brassica* recently has been investigated by Axelsson et al. (2000). They produced synthetic *B. juncea* by doubling the *B. rapa* $\times$ *B. nigra* hybrid. Using nuclear RFLP probes on the synthetic and natural polyploid, they found that its genome has remained essentially unchanged since polyploid formation. The genetic map of the synthetic polyploid could be perfectly integrated with the maps of *B. nigra* and *B. rapa.* Thus, Axelsson et al.

Table 9.2. Types of fragment changes in F_5 progenies of
synthetic *Brassica* polyploids compared with F_2 parents

| | Genome | | | | | |
| | A | | B | | C | |
Polyploid line	Loss	Gain	Loss	Gain	Loss	Gain
A × B	9	13	25	12		
B × A	8	12	14	0		
A × C	7	1			19	4
C × A	15	1			16	5

Source: Adapted from Song et al. (1995).

concluded that whatever was responsible for the results of Song et al. (1995) described above, it was not homoeologous recombination.

Liu et al. (1998b) monitored RFLP fragment profiles in synthetic tetraploids, hexaploids, octoploids, and decaploids in *Aegilops* and *Triticum*. Analyzing the third-and sixth-generation polyploids, they found that (1) rapid genome change was a general phenomenon, (2) the predominant pattern of change was nonrandom sequence elimination from one of the constituent genomes, and (3) that simultaneous loss and appearance of fragments occurs in some polyploids. It is noteworthy that where synthetic polyploids were obtained by crossing species with different ploidal levels, sequence elimination was confined to the lower ploidy parent. Lines with reciprocal parentage did not differ in levels of fragment change. This study involved low-copy noncoding DNA sequences. Liu et al. (1998a) also observed fragment changes in low-copy coding DNA sequences.

Synthetic polyploids also may show the rapid establishment of nucleolar dominance. The nucleolar organizer regions contain the tandemly arrayed genes that encode the precursors of the 18S, 5.8S, and 25S–28S ribosomal RNAs (Givins and Phillips, 1976). Chen et al. (1998) found that in synthetic *Arabidopsis suecica* two generations were required to establish nucleolar dominance in all lines. This species is derived from *A. thaliana* and *Cardaminopsis arenosa*, and the rRNA genes of the former are silent.

The rapid ascendance of nucleolar dominance also has been described in *Cardamine insueta* and *C. schlutzii* (Franzke and Mummenhoff, 1999). The former is a triploid that contains two genomes of *C. rivularis* and one of *C. amara*. The latter is a hexaploid derivative of this triploid that arose via chromosome doubling. Both *C. insueta* and *C. schlutzii* arose early in the twentieth century. In both species, the ITS sequences of rDNA show a very strong bias (87% of the variable nucleotide sequences) toward *C. rivularis*.

Changes in resynthesized polyploids are not restricted to the molecular level. Schranz and Osborn (2000) documented novel flowering time variation in the same synthetic *Brassica napus* lines used by Song et al. (1995), as discussed above. All plants within a lineage should be homozygous, because they were derived from the same plant and extended by self-fertilization. Nevertheless, substantial flowering time variation was observed within lineages and among sublineages that were established in the second generation. Moreover, this variation was highly heritable ($h^2 = 0.60$).

Most synthetic allopolyploids exhibit elevated levels of mortality and developmental aberrations. Thus, it is not surprising to find that newly arisen allopolyploids may be unstable or divergent from their prototypes. Unusual homeotic phenotypes have been observed in *Gilia* (Grant, 1956), *Digitalis* (Schwanitz, 1957), and cotton (Meyer, 1970), flower variegation and tumor formation in *Nicotiana* (Kehr and Smith, 1954; Burns and Gerstel, 1967), and homeotic phenotypes and flower variegation in *Arabidopsis* (Comai et al., 2000). Synthetic allopolyploids are more unstable than synthetic autopolyploids (Allard, 1960).

Why the saltational genome evolution in neoallopolyploids? Comai (2000) proposes that epigenetic interactions between parental genomes are at the heart of it all. This would take the form of altered gene regulation, perhaps due to the activation of transposons. If the genomes have sufficient affinity to allow crossing over, then recombination also could promote rapid evolution.

RECOMBINATIONAL SPECIATION

Once polyploids have evolved, they may diversify into races and species in the same way as diploids do (Stebbins, 1971). However, polyploids with the same number of chromosomes may exchange genes more readily than do diploid species in the same complex, especially if the former have a genome in common. This differential has been observed in *Paeonia* (Stebbins, 1939), *Zauschneria* (Clausen et al, 1945), *Solanum* (Magoon et al., 1962), *Achillea* (Ehrendorfer, 1959), and *Eriophyllum* (Mooring, 2001) among others.

The ability of polyploid species to hybridize and produce fertile hybrids affords the possibility of recombinational speciation. That recombinational speciation may indeed occur at the polyploid level was demonstrated experimentally by Grant (1966b). He crossed *Gilia malior* ($2n = 36$) and *G. modocensis* ($2n = 36$) and obtained a hybrid that was almost completely sterile. Its pollen fertility averaged 2% and its seed fertility less than 1%. After 10 generations of selection for vigor and fertility, Grant obtained a line that was completely fertile with a unique combination of parental attributes and that was intersterile with the parental entities. The chromosome number in the stabilized line was $2n = 38$.

The first report of natural recombinational speciation involving allotetraploid parents recently was made by Ferguson and Sang (2001). Phylogenies of *Adh1* and *Adh2* genes suggest that the widespread *Paeonia officinalis* is derived from *P. peregrina* and a member of the *P. arienta* species assemblage. Each of the putative parental types had two distinct *Adh1* and *Adh2* sequences. The hybrid derivative had three distinct types of *Adh* sequences, two of which are most similar to the two homoeologous *Adh* loci of the *P. arienta* assemblage, and one of which is similar to one of the two *Adh* homoeologues of *P. peregrina*. The other *Adh* homoeologue of this species has been lost. As noted by Ferguson and Sang (2001), recombinational speciation between allotetraploids permits the integration of multiple diploid genomes, which in turn may afford unusual evolutionary potential to the new lineage.

OVERVIEW

Molecular techniques developed in the 1960s and later revealed genetic markers in diploid species that facilitated the identification of genomic constitution of polyploids. These techniques recently have been augmented by "chromosome painting," which allows the visualization of different genomes. The putative parentage of polyploids has been established in tens of genera, and along with it a better understanding of species relationships as a whole.

Several of the more comprehensive investigations have revealed that certain polyploids originated more than once in different locations, in some instances at numerous locations. It also is possible that the same polyploid arises more than once at the same locality. The multiple independent origins of polyploids are contrary to the traditional view that a polyploid species arises only one time at one place. Such origins are important sources of genetic variation for taxa that probably experienced serious bottlenecks at their inception. It is noteworthy that the character of each conspecific polyploid lineage not only is dependent on the nuclear genetic composition of the parental populations, but also may depend on the cytoplasmic identity of the polyploid. Once nuclear–cytoplasmic interactions are better understood, we may find that they play a significant role in determining which polyploids succeed and which do not.

By virtue of possessing two or more divergent genomes, allopolyploids have higher levels of heterozygosity than their diploid prototypes. The more disparate the genomes and the greater the number of different genomes the greater is the level of heterozygosity. Heterozygosity also tends to increase as the ploidal level increases. Fixed heterozygosity is the hallmark of allopolyploidy. It is thought to be one of the prime factors behind the success of allopolyploids.

The traditional view that the genomes of allopolyploids are independent entities, each carrying on as before they were joined, has been found wanting. Extensive intergenomic chromosome reorganization via translocations has been demonstrated in several species. This renders the genomes of allopolyploids more structurally divergent than the same genomes in the diploid progenitors. Intergenomic sequence transfer also occurs and may be accompanied by concerted evolution. Gene silencing is another manifestation of intergenomic interaction. It is most common regarding rDNA genes. In some polyploids, gene silencing is so extensive that inheritance is disomic.

In some polyploid lineages, gene silencing has been accompanied by enhanced regulation of chromosome pairing such that only bivalents are formed or multivalent formation at zygotene is corrected by pachytene. The diploid appearance of these species is only belied by the extensive redundancy of genetic markers.

The aforementioned changes may be accompanied by alterations in genome size. Some allopolyploids have significantly less DNA than the sum of their progenitors. Whether there is equal loss from each genome remains to be determined.

The picture that emerges for polyploids is one of considerable genomic change. Indeed, alterations may occur within the first few generations of their existence. They are manifested at the molecular level and at the organismic level. Epigenetic phenomena or recombination between homoeologous chromosomes could generate the instability allowing rapid evolution.

Bibliography

Abdel-Hameed, F. 1971. Cytogenetic studies in *Clarkia*, section *Primigenia*. V. Interspecific hybridization between *C. amoena huntiana* and *C. lassenensis*. Evolution 25: 347–355.

Ahn, S., and S.D. Tanskley. 1993. Comparative linkage maps of the rice and maize genomes. Proc. Natl. Acad. Sci. USA 90: 7980–7984.

Albuzio, A., P. Spettoli, and G. Cacco. 1978. Changes in gene expression from diploid to autotetraploid status of *Lycopersicon esculentum*. Plant Physiol. 44: 77–80.

Alexander, D.E. 1960. Performance of genetically induced corn tetraploids. Am. Seed Trade Assoc. 15: 68–74.

Allard, R.W. 1960. Principles of Plant Breeding. Wiley, New York.

Allen, G.A. 2001. Hybrid speciation in *Erythronium* (Liliaceae): a new allotetraploid species from Washington State. Syst. Bot. 26: 263–272.

Allen, G.A., and C.L. Eccleston. 1998. Genetic resemblance of allotetraploid *Aster ascendens* to its diploid progenitors *Aster falcatus* and *Aster occidentalis*. Can. J. Bot. 76: 338–344.

Al-Sahael, Y.A., and A.S. Larik. 1985. Genetic factors determining plasticity in flax genotrophs. Can. J. Genet. Cytol. 27: 272–275.

Anderson, E. 1945. What is *Zea mays?* A report of progress. Chron. Bot. 9: 88–92.

Anderson, L.B. 1971. A study of some seedling characteristics and the effects of competition on seedlings in diploid and tetraploid red clover (*Trifolium pratense* L.). N.Z. J. Agric. Res. 14: 563–571.

Arbi, N., D. Smith, E.T. Bingham, and R.M. Soberalske. 1978. Herbage yields and levels of N and IVDDM from five alfalfa strains of different ploidy levels. Agron. J. 70: 873–875.

Arft, A.M., and T.A. Ranker. 1998. Allopolyploid origin and population genetics of the rare orchid *Spiranthes diluvialis*. Am. J. Bot. 85: 110–122.

Asker, S.E., and L. Jerling. 1992. Apomixis in Plants. CRC Press, London.

Attia, T., and G. Robbelen. 1986a. Meiotic pairing in haploids and amphidiploids of spontaneous versus synthetic origin rape, *Brassica napus* L. Can. J. Genet. Cytol. 28: 330–334.

Attia, T., and G. Robbelen. 1986b. Cytogenetic relationships within cultivated *Brassica* analyzed in amphihaploids from the three diploid ancestors. Can. J. Genet. Cytol. 28: 323–329.

Auger, D.L., K.J. Newton, and J.A. Birchler. 2001. Nuclear gene dosage effects upon the expression of maize mitochondrial genes. Genetics 157: 1711–1721.

Aung, T., and G.M. Evans. 1987. Segregation of isozyme markers and meiotic pairing control genes in *Lolium*. Heredity 59: 129–134.

Avdulov, N.P. 1931. Karyo-systematische Untersuchungen der Famalie Gramineen. Bull. Appl. Bot. Genet. Plant Breed. 44 (Suppl. 4): 1–428.

Avery, G.S., and L. Pottorf. 1945. Polyploidy, auxin, and nitrogen in green plants. Am. J. Bot. 32: 669–671.

Axelsson, T., C.M. Bowman, A.G. Sharpe, D.J. Lydiate, and V. Langercrantz. 2000. Amphiploid *Brassica juncea* contains conserved progenitor genomes. Genome 43: 679–688.

Babcock, E.B. 1947. The genus *Crepis,* vols. I and II. University of California Publications on Botany, Berkeley.

Badaeva, E., B. Friebe, and B.S. Gill. 1996a. Genome differentiation in *Aegilops*. 1. Distribution of highly repetitive DNA sequences on chromosomes of diploid species. Genome 39: 293–306.

Badaeva, E., B. Friebe, and B.S. Gill. 1996b. Genome differentiation in *Aegilops*. 2. Physical mapping of 5S and 18S-26S ribosomal RNA gene families in diploid species. Genome 39: 1150–1158.

Bailey, R.J., H. Rees, and L.M. Jones. 1976. Interchange heterozygotes versus homozygotes. Heredity 37: 109–112.

Baldwin, B.G. 1993. Molecular phylogenetics of *Calycadenia* (Compositae) based on its sequences of nuclear ribosomal DNA: chromosomal and morphological evolution reexamined. Am. J. Bot. 80: 222–238.

Baldwin, B.G. 1997. Adaptive radiation of the Hawaiian silversword alliance: congruence and conflict of phylogenetic evidence from molecular and non-molecular investigations. Pages 103–128 in T.J. Givnish and K.J. Sytsma, eds., Molecular Evolution and Adaptive Radiation. Cambridge University Press, Cambridge.

Baldwin, B.G., and S. Markos. 1998. Phylogenetic utility of the external transcribed spacer (ETS) of 18S-26S rDNA: congruence of ETS and ITS trees of *Calycadenia* (Compositae). Mol. Phylogenet. Evol. 10: 449–463.

Baldwin, J.T., Jr. 1941. *Galax:* the genus and its chromosomes. J. Hered. 32: 249–254.

Balint-Kurti, P.J., M.S. Dixon, D.A. Jones., K.A. Norcott, and J.D.G. Jones. 1994. RFLP linkage analysis of the Cf-4 and Cf-9 genes for resistance to *Cladosporium fulvum* in tomato. Theor. Appl. Genet. 88: 691–700.

Banerjee, P.K. 1968. Variation of chemical content with change in chromosome number. Pages 324–331 in A.K. Sharma and A. Sharma, eds., Proceedings of the International Seminar on the Chromosome Nucleus (Suppl.). University of Calcutta, Calcutta.

Barber, J.C., J. Francisco-Ortega, A Santos-Guerra, A. Marrero, and R.K. Jansen. 2000. Evolution of endemic *Sideritis* (Lamiaceae) in Macaronesia: insights from a chloroplast DNA restriction site analysis. Syst. Bot. 25: 633–647.

Barcaccia, G., E. Albertini, D. Rosellini, S. Tavoletti, and F. Veronesi. 2000. Inheritance and mapping of 2n-egg production in diploid alfalfa. Genome 43: 528–537.

Barton, N.H. 1979. The dynamics of hybrid zones. Heredity 43: 341–359.

Barton, N.H., and G.M. Hewitt. 1985. Analysis of hybrid zones. Annu. Rev. Ecol. Syst. 16: 113–148.

Batra, S. 1952. Induced tetraploidy in muskmelons. J. Hered. 43: 141–148.

Bayer, R.J. 1997. *Antennaria rosea* (Asteraceae)—a model group for the study of the evolution of agamic complexes. Opera Bot. 132: 53–65.

Bayer, R.J., and G.L. Stebbins. 1987. Chromosome numbers, patterns of distribution, and apomixis in *Antennaria* (Asteraceae: Inuleae). Syst. Bot. 12: 305–319.

Bazzaz, F.A., D.A. Levin, and M. Levy. 1983. The effect of chromosome doubling on photosynthetic rates in *Phlox*. Photosynthetica 7: 89–92.

Belling, J. 1925. The origin of chromosomal mutations in *Uvularia*. J. Genet. 15: 245–266.

Beltran, I.C., and S.H, James. 1974. Complex hybridity in *Isotoma petraea*. IV. Heterosis in interpopulation hybrids. Austral. J. Bot. 22: 251–264.

Benavente, E., and J. Orellana. 1991. Chromosome differentiation and pairing behavior of polyploids: an assessment of preferential metaphase I associations in colchicine-induced autotetraploid hybrids within the genus *Secale*. Genetica 128: 433–442.

Benavente, E., J. Orellana, and B. Fernández-Calvin. 1998. Comparative analysis of the meiotic effects of wheat *ph1b* and *ph2b* mutations in wheat × rye hybrids. Theor. Appl. Genet. 96: 1200–1204.

Bennett, J.H. 1954. Panmixia with tetrasomic and hexasomic inheritance. Genetics 39: 150–158.

Bennett, M.D. 1972. Nuclear DNA content and minimum generation time. Proc. R. Soc. Lond. B. 181: 109–135.

Bennett, M.D. 1976. DNA amount, latitude, and crop plant distribution. Environ. Exp. Bot. 16: 93–108.

Bennett, M.D. 1985. Intraspecific variation in DNA amount and the nucleotype dimension in plant genetics. Pages 283–302 in M. Freeling, ed., Plant Genetics. Liss, New York.

Bennett, M.D. 1987. Variation in genomic form in plants and its ecological implications. New Phytol. 106 (Suppl.): 177–200.

Bennett, M.D., and I.J. Leitch. 1997. Nuclear DNA amounts in angiosperms—583 new estimates. Ann. Bot. 80: 169–196.

Bennett, M.D., and J.B. Smith. 1976. Nuclear DNA amounts in angiosperms. Phil. Trans. R. Soc. Lond. B. 274: 227–274.

Bennett, M.D., J.B. Smith, and J.S. Heslop-Harrison. 1982. Nuclear DNA amounts in angiosperms. Proc. R. Soc. Lond. B. 216: 179–199.

Bennetzen, J.L. 1996. The contribution of retroelements to plant genome organization, function and evolution. Trends Microbiol. 4: 347–353.

Bennetzen, J.L., and E.A. Kellogg. 1997. Do plants have a one-way ticket to genome obesity? Plant Cell 9: 1509–1514.

Bentzer, B. 1972a. Structural chromosome polymorphism in the diploid *Leopoldia weissii* Freyn ex Heldr. (Liliaceae). Bot. Not. 125: 180–185.

Bentzer, B. 1972b. Variation in the chromosome complement of *Leopoldia comosa* (L.) Parl. (Liliaceae) in the Aegean (Greece). Bot. Not. 125: 406–418.

Bernstrom, P. 1950. Cleisto- and chasmogamic seed setting in di- and tetraploid *Lamium amplexicaule*. Hereditas 36: 492–506.

Bever, J.D., and F. Felber. 1992. The theoretical population genetics of autopolyploidy. Pages 185–217 in J. Antonovics and D. Futuyama, eds., Oxford Surveys in Evolutionary Biology, vol. 8. Oxford University Press, New York.

Beysel, D. 1957. Assimilations und Atmungmessungen an diploiden und polyploiden Zuckerrüben. Züchter 27: 261–272.

Bharathan, G., G. Lambert, and D.W. Galbraith. 1994. Nuclear DNA content of monocotyledons and related taxa. Am. J. Bot. 81: 381–386.

Bhatt, B., and M.R. Heble. 1978. Solasodine content in fruits of spiny and mutant tetraploids of *Solanum khasianum* Clarke. Environ. Exp. Bot. 18: 127–130.

Bingham, E.T. 1980. Maximizing heterozygosity in polyploids. Pages 471–490 in W.H. Lewis, ed., Polyploidy—Biological Relevance. Plenum, New York.

Bingham, E.T., R.W. Groose, D.R. Woodfield, and K.K. Kidwell. 1994. Complementary gene interactions in alfalfa are greater in autotetraploids than diploids. Crop Sci. 34: 823–829.

Birari, S.P. 1980. Apomixis and sexuality in *Themeda* Foressk. at different ploidy levels (Gramineae). Genetica 54: 133–139.

Bjurman, B. 1959. The photosynthesis in diploid and tetraploid *Ribes satigrum*. Physiol. Plant. 12: 183–187.

Blake, N., B.R. Lehfeldt, M. Lavin, and L.E. Talbert. 1999. Phylogenetic reconstruction based on low copy DNA sequence data in an allopolyploid: the B genome of wheat. Genome 42: 351–360.

Bland, M.M., D.F. Matzinger, and C.S. Levings. 1985. Comparison of the mitochondrial genome of *Nicotiana tabacum* with its progenitor species. Theor. Appl. Genet. 69: 535–541.

Bloom, W.L. 1974. Origin of reciprocal translocations and their effect in *Clarkia speciosa*. Chromosoma 49: 61–76.

Bloom, W.L. 1976. Multivariate analysis of the introgressive replacement of *Clarkia nitens* by *Clarkia speciosa polyantha*. Evolution 30: 412–424.

Bloom, W.L., and H. Lewis. 1972. Interchanges and interpopulational gene exchange in *Clarkia speciosa*. Chromosomes Today 3: 268–284.

Bonierbale, M.W., R.L. Plaisted, and S.D. Tanksley. 1988. RFLP maps based on a common set of clones reveal modes of chromosomal evolution in potato and tomato. Genetics 120: 1095–1103.

Borrill, M., and R. Lindner. 1971. Diploid-tetraploid sympatry in *Dactylis* (Gramineae). New Phytol. 70: 1111–1124.

Bose, R.B., and J.K. Choudhury. 1962. A comparative study of the cytotaxonomy, palynology and physiology of diploid plants from *Ocimum kilimandscharicum* Guerke and their yield of raw material and volatile contents. Caryologia 15: 435–453.

Bothmer, R., von. 1970. Cytological studies in *Allium*. Chromosome numbers and morphology in sect. *Allium* from Greece. Bot. Not. 123: 519–551.

Bouharmont, J., and F. Mace. 1972. Valeur competitive des plantes autopolyploides d'*Arabidopsis thaliana*. Can. J. Genet. Cytol. 14: 257–263.

Brammel, R.A., and J.C. Semple. 1990. The cytotaxonomy of *Solidago nemoralis* (Compositae: Asteraceae). Can. J. Bot. 68: 2065–2069.

Brandham, P.E. 1974. Interchange and inversion polymorphism among populations of *Haworthia reinwardtii* var. *chalumnensis*. Chromosoma 47: 85–108.

Brandham, P.E. 1983. Evolution in a stable chromosome system. Pages 251–260 in P.E. Brandham and M.D. Bennett, eds., Kew Chromosome Conference II. Allen and Unwin, Boston.

Brandham, P.E., and M.J. Doherty. 1998. Genome size variation in the Aloaceae, an angiosperm family displaying karyotypic orthoselection. Ann. Bot. 82 (Suppl. A): 67–73.

Breiman, A., and Grauer, D. 1995. Wheat evolution. Israel J. Plant Sci. 43: 85–98.

Bremer, G., and D.E. Bremer-Reinders. 1954. Breeding of tetraploid rye in the Netherlands. I. Methods and cytological investigations. Euphytica 3: 49–63.

Bretagnolle, F. 2001. Pollen production and spontaneous polyploidization in diploid populations of *Anthoxanthum odoratum*. Biol. J. Linn. Soc. 72: 241–247.

Bretagnolle, F., and J.D. Thompson. 1995. Gametes with the somatic chromosome number: mechanisms of their formation and role in the evolution of polyploid plants. New Phytol. 129: 1–22.

Brewbaker, J.L. 1958. Self-compatibility in tetraploid strains of *Trifolium hybridum* L. Hereditas 44: 547–553.

Brighton, C.A. 1976. Cytological problems in the genus *Crocus* (Iridaceae). I. *Crocus vernus* aggregate. Kew Bull. 31: 33–46.

Brighton, C.A. 1978. Cytological problems in the genus *Crocus* (Iridaceae): II. *Crocus cancellatus* aggregate. Kew Bull. 32: 33–44.

Brighton, C.A., B. Mathew, and P. Rudall. 1983. A detailed study of *Crocus speciosus* and its ally *C. pulchellus* (Iridaceae). Plant Syst. Evol. 142: 187–206.

Brink, R.A., and D.C. Cooper. 1947. The endosperm in seed development. Bot. Rev. 132: 432–541.

Brochmann, C.L. Borgen, and O.E. Stabbetorp. 2000. Multiple diploid hybrid speciation of the Canary Island endemic *Argyranthemum sundingii* (Asteraceae). Plant Syst. Evol. 220: 77–92.

Brochmann, C., and R. Elven. 1992. Ecological and genetic consequences of polyploidy in arctic *Draba* (Brassicaceae). Evol. Trends Plants 6: 111–124.

Brochmann, C., P.S. Soltis, and D.E. Soltis. 1992a. Recurrent formation and polyphyly of Nordic polyploids in *Draba* (Brassicaceae). Am. J. Bot. 79: 673–688.

Brochmann, C., P.S. Soltis, and D.E. Soltis. 1992b. Multiple origins of the octoploid Scandinavian endemic *Draba cacuminum*—electrophoretic and morphological evidence. Nordic J. Bot. 12: 257–272.

Brochmann, C., T. Nillson, and T.M. Gabrielsen. 1996. A classic example of allopolyploid speciation re-examined using RAPD markers and nucleotide sequences: *Saxifraga osloensis* (Saxifragaceae). Symb. Bot. Upsala 31: 75–89.

Brubaker, C.L., A.H. Paterson, and J.F. Wendel. 1999. Comparative genetic mapping of allotetraploid cotton and its diploid progenitors. Genome 42: 184–203.

Brunsberg, K. 1997. Biosystematics of the *Lathyros pratensis* complex. Opera Bot. 42: 1–78.

Burdon, J.J., and D.R. Marshall. 1981. Intra- and interspecific diversity in the disease response of *Glycine* species to the leaf-rust fungus *Phakospora pachyrhizi*. J. Ecol. 69: 381–390.

Burns, J., and D. Gerstel. 1967. Flower color variegation and instability of a block of heterochromatin in *Nicotiana*. Genetics 57: 155–167.

Burton, T.L., and B.C. Husband. 1999. Population cytotype structure in the polyploid *Galax urceolata* (Diapensiaceae). Heredity 82: 381–390.

Busey, P., R.M. Giblin-Davis, and B.J. Center. 1993. Resistance in *Stentaphrum* to the stinging nematode. Crop Sci. 33: 1066–1070.

Byrne, M.C., C.J. Nelson, and D.D. Randall. 1981. Ploidy effects on anatomy and gas exchange of tall fescue. Plant Physiol. 68: 891–893.

Cacco, G., G. Ferrari, and G.C. Lucci. 1976. Uptake efficiency of roots of in plants at different ploidy levels. J. Agric. Sci. 87: 585–589.

Caceres, M.E., C. De Pce, G.T. Scarascia Mugnozza, P. Kotsonis, M. Ceccarelli, and

P.G. Cionini. 1998. Genome size variations within *Dasypyrum villosum:* correlation with chromosomal traits, environmental factors and plant phenotypic characteristics and behaviour in reproduction. Theor. App. Genet. 96: 559–567.

Calderini, O., and A. Mariani. 1997. Increasing 2*n* gamete production in diploid alfalfa by cycles of phenotypic recurrent selection. Euphytica 93: 113–118.

Carr, B.L., D.P. Gregory, P.H. Raven, and W. Tai. 1986a. Experimental hybridization and chromosome diversity within *Gaura* sect. *Gaura* (Onagraceae). Syst. Bot. 11: 98–111.

Carr, B.L., D.P. Gregory, P.H. Raven, and W. Tai. 1986b. Experimental hybridization and chromosome diversity within *Gaura* sect. *Stipogaura* (Onagraceae). Am. J. Bot. 73: 1144–1156.

Carr, B.L., D.P. Gregory, P.H. Raven, and W. Tai. 1988. Experimental hybridization and chromosome diversity within *Gaura* sect. *Pterogaura* (Onagraceae). Syst. Bot. 13: 324–335.

Carr, G.D. 1975. Chromosome evolution and aneuploid reduction in *Calycadenia pauciflora* (Asteraceae). Evolution 29: 681–699.

Carr, G.D. 1977. A cytological conspectus of the genus *Calycadenia* (Asteraceae): an example of contrasting modes of evolution. Am. J. Bot. 64: 694–703.

Carr, G.D. 1981. Experimental evidence for saltational chromosome evolution in *Calycadenia pauciflora* Gray (Asteraceae). Heredity 45: 107–112.

Carr, G.D. 1998. Chromosome evolution and speciation in Hawaiian flowering plants. Pages 5–47 in T.F. Stuessy and M. Ono, eds., Evolution and Speciation of Island Plants. Cambridge University Press, New York.

Carr, G.D., and R.L. Carr. 1983. Chromosome races and structural heterozygosity in *Calycadenia ciliosa* Greene (Asteraceae). Am. J. Bot. 70: 744–755.

Carr, G.D., and R.L. Carr. 2000. A new chromosome race of *Calycadenia pauciflora* (Asteraceae: Heliantheae-Madiinae) from Butte County, California. Am. J. Bot. 87: 1459–1465.

Carr, G.D., and D.W. Kyhos. 1981. Adaptive radiation in the Hawaiian silversword alliance (Compositae-Madiinae). I. Cytogenetics of spontaneous hybrids. Evolution 35: 543–556.

Carr, G.D., and D.W. Kyhos. 1986. Adaptive radiation in the Hawaiian silversword alliance (Compositae-Madiinae). II. Cytogenetics of artificial and natural hybrids. Evolution 40: 959–976.

Cavalier-Smith, T. 1978. Nuclear volume control by nucleoskeleton DNA, selection for cell volume and cell growth rate, and the solution of the DNA C-value paradox. J. Cell Sci. 34: 247–278.

Cavallini, A., L. Natali, G. Cionini, and D. Gennai. 1993. Nuclear DNA variability within *Pisum sativum* (Leguminosae): nucleotypic effects on plant growth. Heredity 70: 561–565.

Cavanah, J.A., and D.E. Alexander. 1963. Survival of tetraploid maize in 2*n*-4*n* plantings. Crop Sci. 3: 329–331.

Cave, M.S., and L. Constance. 1947. Chromosome numbers in the Hydrophyllaceae. III. Univ. Calif. Publ. Bot. 18: 449–465.

Cave, M.S., and L. Constance. 1950. Chromosome numbers in the Hydrophyllaceae. IV. Univ. Calif. Publ. Bot. 23: 363–382.

Ceccarelli, M., E. Falistocco, and P.G. Cionini. 1992. Variation of genome size and organization within hexaploid *Festuca arundinacea*. Theor. Appl. Genet. 83: 273–278.

Ceccarelli, M., S. Minelli, F. Maggini, and P.G. Cionni. 1995. Genome size variation in *Vicia faba*. Heredity 74: 180–187.

Celarier, R.P. 1956. Additional evidence of five as the basic chromosome number of the Andropogoneae. Rhodora 58: 135–148.

Cerbah, M., J. Coulaud, S.C. Brown, and S. Siljak-Yakovlev. 1999. Evolutionary DNA variation in the genus *Hypochaeris*. Heredity 82: 261–266.

Cerbah, M., J. Coulaud, and S. Siljak-Yakovlev. 1998a. rDNA organization and evolutionary relationships in the genus *Hypochaeris* (Asteraceae). J. Hered. 89: 312–318.

Cerbah, M., T. Souza-Chies, M.F. Joubier, B. Lejeune, and S. Siljak-Yakovlev. 1998b. Molecular phylogeny of the genus *Hypochaeris* using internal transcribed spacers of nuclear rDNA: inference for chromosomal evolution. Mol. Biol. Evol. 15: 345–354.

Chandler, J.M., C.-C. Jan, and B.H. Beard. 1986. Chromosomal differentiation among annual *Helianthus* species. Syst. Bot. 11: 354–371.

Chao, C.Y. 1974. Megasporogenesis and megagametogenesis in *Paspalum commersonii* and *P. longiflorum* at two ploidal levels. Bot. Not. 127: 267–275.

Charlesworth, B. 1987. The population biology of transposable elements. Trends Ecol. Evol. 2: 21–23.

Charpentier, A., M. Feldman, and Y. Cauderon. 1988. The effect of different *Agropyron elongatum* chromosomes on pairing in *Agropyron*-common wheat hybrids. Genome 30: 978–983.

Chatterje, R., and G. Jenkins. 1993. Meiotic chromosome interactions in inbred autotetraploid rye (*Secale cereale*). Genome 36: 131–138.

Chauhan, K.P.S., and M.S. Swaminathan. 1984. Cytological effects of aging in seeds. Genetica 64: 69–76.

Chen, B.Y., W.K. Heenen, and V. Simonson. 1989. Comparative and genetic studies of isozymes in resynthesized and cultivated *Brassica napus* L., *B. campestris* L., and *B. alboglabra* Bailey. Theor. Appl. Genet. 77: 673–679.

Chen, S.L., and P.S. Tang. 1945. Studies on colchicine-induced autotetraploid barley. III. Physiological studies. Am. J. Bot. 32: 177–179.

Chen, Z.F., L. Comai, and C.S. Pikaard. 1998. Gene dosage and stochastic effects determine the severity and direction of uniparental ribosomal RNA gene silencing (nucleolar dominance) in *Arabidopsis* allopolyploids. Proc. Natl. Acad. Sci. USA 95: 14891–14896.

Chetelat, R.T., and V. Meglic. 2000. Molecular mapping of chromosome segments introgressed from *Solanum lycopersicoides* into cultivated tomato (*Lycopersicon esculentum*). Theor. Appl. Genet. 100: 232–241.

Chetelat, R.T., V. Meglic, and P. Cisneros. 2000. A genetic map of tomato based on BC_1 *Lycopersicon esculentum* × *Solanum lycopersicoides* reveals overall synteny but suppressed recombination between these homoeologous genomes. Genetics 154: 857–867.

Chooi, W.Y. 1971. Variation in nuclear DNA content in the genus *Vicia*. Genetics 68: 195–211.

Choudhury, J.B., B.S. Ghai, and P.K. Sareen. 1968. Studies on the cytology and fertility in induced polyploids of the self-incompatible *Brassica campestris* var. Brown Sarson. Cytologia 33: 269–275.

Chu, Y.E. 1972. Genetic bases, classification and origin of reproductive barriers in *Oryza* species. Bot. Bull. Acad. Sin. 13: 47–66.

Civardi, L., Y. Xia, K.J. Edwards, P.S. Schnable, and B.J. Nicklau. 1994. The relationship between the genetic and physical distances in the cloned *al1-sh2* interval of the *Zea mays* L. genome. Proc. Natl. Acad. Sci. USA 91: 8268–8272.

Clausen, J., D.D. Keck, and W.H. Hiesey. 1945. Experimental studies on the nature of species. II. Plant evolution through amphidiploidy and autopolyploidy with examples from the Madiinae. Publ. 564. Carnegie Institute, Washington, DC.

Cleland, R.E. 1972. *Oenothera:* Cytogenetics and Evolution. Academic Press, New York.

Comai, L. 2000. Genetic and epigenetic interactions in allopolyploid plants. Plant Mol. Biol. 43: 387–399.

Comai, L., A.P. Tyagi, K. Winter, R. Holmes-Davis, S. Reynolds, Y. Stevens, and B. Byers. 2000. Phenotypic instability and gene silencing in newly formed *Arabidopsis* allotetraploids. Plant Cell 12: 1551–1567.

Constance, L. 1963. Chromosome number and classification in the Hydrophyllaceae. Brittonia 15: 273–285.

Cook, L.M., P.S. Soltis, S.J. Brunsfeld, and D.E. Soltis. 1998. Multiple independent formations of *Tragopogon* tetraploids (Asteraceae): evidence from RAPD markers. Mol. Ecol. 7: 1293–1302.

Coulthart, M., and K.E. Denford. 1982. Isozyme studies in *Brassica.* 1. Electrophoretic techniques for leaf enzymes and comparison of *B. napus, B. campestris,* and *B. oleracea* using phosphoglucoisomerase. Can. J. Plant Sci. 62: 621–630.

Cox, A.V., G.J. Abdelnour, M.D. Bennett, and I.J. Leitch. 1998. Genome size and karyotype evolution in the slipper orchids (Cypripedioidae: Orchidaceae). Am. J. Bot. 85: 681–687.

Coyne, J.A. 1996. Speciation in action. Science 272: 700–701.

Crane, M.B., and D. Lewis. 1942. Genetical studies in pears. III. Incompatibility and sterility. J. Genet. 43: 31–43.

Crawford, D.J., and E.B. Smith. 1982. Allozyme variation in *Coreopsis nucensoides* and *C. nuecensis* (Compositae), a progenitor-derivative species pair. Evolution 36: 379–386.

Crawford, T.J. 1984. What is a population? Pages 135–173 in B. Shorrocks, ed., Evolutionary Ecology. Blackwell, London.

Cronn, R.C., R.L. Smith, and J.F. Wendel. 1999. Duplicated genes evolve independently after polyploid formation in cotton. Proc. Natl. Acad. Sci. USA 96: 14406–14411.

Cros, J. M.C. Combes, N. Chabrillange, C. Duperray, A. Monnot des Angles, and S. Hamon. 1995. Nuclear DNA content in the subgenus *Coffea* (Rubiaceae): inter- and intraspecific variation in African species. Can. J. Bot. 73: 14–20.

Crow, J.F., and M. Kimura. 1970. An Introduction to Population Genetics Theory. Harper and Row, New York.

Crowley, J.G., and H. Rees. 1968. Fertility and selection in tetraploid *Lolium.* Chromosoma 24: 300–308.

Cukrova, V., and N. Avratocscukova. 1968. Photosynthetic activity, chlorophyll content, and stomatal characteristics in diploid and tetraploid types of *Datura stramonium.* L. Photosynthetica 2: 227–228.

Cullis, C.A. 1976. Environmentally induced changes in ribosomal RNA cistron number in flax. Heredity 36: 73–79.

Cullis, C.A. 1979. Quantitative variation of ribosomal RNA genes in flax genotrophs. Heredity 42: 237–246.

Cullis, C.A. 1990. DNA rearrangements in response to environmental stress. Adv. Genet. 28: 73–97.

Cullis, C., and D.R. Davies. 1974. Ribosomal RNA cistron number in a polyploid series of plants. Chromosoma 46: 23–28.

Cuñado, N., S. Callejas, M.J. García, A. Fernández, and J.L. Santos. 1996a. The pattern of zygotene and pachytene pairing in allotetraploid *Aegilops* species sharing the U genome. Theor. Appl. Genet. 93: 1152–1155.

Cuñado, N., S. Callejas, M.J. García, A. Fernández, and J.L. Santos. 1996b. The pattern of zygotene and pachytene pairing in allotetraploid *Aegilops* species sharing the D genome. Theor. Appl. Genet. 93: 1175–1179.

Darlington, C.D. 1932. The control of the chromosomes by the genotype and its bearing on some evolutionary problems. Am. Nat. 66: 25–51.

Darlington, C.D. 1936. The limitation of crossing over in *Oenothera*. J. Genet. 32: 343–352.

Darlington, C.D. 1937. Recent Advances in Cytology, 2nd ed. Churchill, London.

Darlington, C.D. 1958. The Evolution of Genetic Systems, 2nd ed. Basic Books, New York.

Darlington, C.D., and L.F. La Cour. 1950. Hybridity selection in *Campanula*. Heredity 4: 217–248.

Dass, H., and H. Nybom. 1967. The relationships between *Brassica nigra, B. campestris, B. oleracea,* and their amphidiploid hybrids studied by means of numerical taxonomy. Can. J. Genet. Cytol. 9: 880–890.

Datta, R.M. 1963. Investigations on the autotetraploids of the cultivated and wild types of jute (*Corchorus olitorius* and *C. capsularis* L.). Züchter 33: 17–33.

Davies, A., G. Jenkins, and H. Rees. 1990. Diploidization of *Lotus corniculatus* L. (Fabaceae) by elimination of multivalents. Chromosoma 99: 289–295.

Day, A. 1965. The evolution of a pair of sibling allotetraploid species of cobwebby gilias (Polemoniaceae). Aliso 6: 25–75.

De Haan, A., N.O. Maceira, R. Lumaret, and J. Delay. 1992. Production of 2*n* gametes in diploid subspecies of *Dactylis glomerata* L. 2. Occurrence and frequency of 2*n* eggs. Ann. Bot. 69: 345–350.

Delseny, M.Y., J.M. McGrath, P. This, A.M. Chevre, and C.F. Queros. 1990. Ribosomal RNA genes in diploid and amphidiploid *Brassica* and related species: organization, polymorphism, and evolution. Genome 33: 733–744.

de Nettancourt, D. 2001. Incompatibility and Incongruity in Wild and Cultivated Plants. Springer-Verlag, Berlin.

de Nettancourt, D., F. Saccardo, U. Lanieri, E. Capaccio, M. Westerhof, and R. Ecochard. 1974. Self-compatibility in a spontaneous tetraploid of *Lycopersicon peruvianum* Mil. Pages 77–84 in Polyploidy and Induced Mutations in Plant Breeding. International Atomic Energy Agency, Vienna.

deWet, J.M.J. 1980. Origins of polyploids. Pages 3–16 in W.H. Lewis, ed., Polyploidy—Biological Relevance. Plenum, New York.

Devos, K.J., M.D. Atkinson, C.S. Chino, R.L. Harcourt, R.M.T. Koebner, C.J. Liu, P. Masojc, D.X. Xie, and M.D. Gale. 1993. Chromosomal rearrangements in the rye genome relative to that of wheat. Theor. Appl. Genet. 85: 673–680.

Devos, K.M., and M.D. Gale. 2000. Genome relationships: the grass model in current research. Plant Cell 12: 637–646.

Dewey, D.R. 1969. Inbreeding depression in diploid and induced-autotetraploid crested wheatgrass. Crop Sci. 9: 592–595.

Dhillon, S.S. 1988. DNA analysis during growth and development. Pages 265–274 in J.W. Hanover and D.E. Keathley, eds., Genetic Manipulation of Woody Plants. Plenum Press, New York.

Diaz Lifante, Z. 1996. A karyological study of *Asphodelus* L. (Asphodeliaceae) from the western Mediterranean. Bot. J. Linn. Soc. 121: 285–344.

Dietrich, W. 1978. The South American species of *Oenothera* sect. *Oenothera* (*Raimannia, Renneria:* Onagraceae). Ann. Missouri Bot. Gard. 64: 425–626.

Dijkstra, H., and G.J. Speckmann. 1980. Autotetraploidy in caraway (*Carum carvi* L.) for the increase of the aetheric oil content of the seed. Euphytica 29: 89–96.

Dimitrova, D., and J. Greilhuber. 2000. Karyotype and DNA content in ten species of *Crepis* (Asteraceae) distributed in Bulgaria. Bot. J. Linn. Soc. 132: 281–297.

Dnyansager, V.R., and I.V. Sudhakaran. 1970. Induced tetraploidy in *Vinca rosea* L. Cytologia 35: 227–241.

Doolittle, W., and C. Sapeinza. 1980. Selfish genes, the phenotypic paradigm and genome evolution. Nature 284: 601–603.

Dooner, H.K., and I.M. Martinez-Ferez. 1997. Recombination occurs uniformly within the *bronze* gene, a meiotic recombination hotspot in the maize genome. Plant Cell 9: 1633–1646.

Döring, H.-P., and P. Salinger. 1986. Molecular genetics of transposable elements in plants. Annu. Rev. Genet. 20: 175–200.

Dover, G. 1982. Molecular drive: a cohesive model of species evolution. Nature 299: 111–117.

Dowrick, G.J., and A.S. El-Bayoumi. 1969. Nucleic acid content and chromosome morphology in *Chrysanthemum*. Genet. Res. 13: 241–250.

Doyle, J.J., J.L. Doyle, A.H.D. Brown, and J.P. Grace. 1990. Multiple origins of polyploids in the *Glycine tabacina* complex inferred from chloroplast DNA polymorphism. Proc. Natl. Acad. Sci. USA 87: 714–717.

Doyle, J.J., J.L. Doyle, and A.H.D. Brown. 1999. Origins, colonization, and lineage recombination in a widespread perennial soybean polyploid complex. Proc. Natl. Acad. Sci. USA 96: 10741–10745.

Dubcovsky, J., and J. Dvořák. 1995. Ribosomal RNA multigene loci: nomads of the Triticeae genomes. Genetics 140: 1367–1377.

Dubcovsky, J., M.-C. Liu, G.-Y. Zhong, R. Bransteitter, A. Desai, A. Kilian, A. Kleinhofs, and J. Dvořák. 1996. Genetic map of diploid wheat *Triticum monococcum* L. and its comparision with maps of *Hordeum vulgare* L. Genetics 143: 983–999.

Dudley, J.W., and D.E. Alexander. 1969. Performance of advanced generations of autotetraploid maize (*Zea mays* L.) synthetics. Crop Sci. 9: 613–615.

Dunbier, M.W., D.L. Eskew, E.T. Bingham, and L.E. Schrader. 1975. Performance of genetically comparable diploid and tetraploid alfalfa: agronomic and physiological parameters. Crop Sci. 15: 211–214.

Durrant, A. 1981. Unstable genotypes. Philos. Trans. R. Soc. Lond. B. 292: 467–474.

Dvořák, J. 1987. The chromosomal distribution of genes in diploid *Elytrigia elongata* that promote or suppress pairing of wheat homoeologous chromosomes. Genome 29: 34–40.

Dvořák, J., and D.B. Fowler. 1978. Cold hardiness of triticale and tetraploid rye. Crop Sci. 18: 477–478.

Eagles, C.F., and O.B. Othman. 1978. Physiological studies of a hybrid between populations of *Dactylis glomerata* from contrasting climatic regions. I. Interpopulation differences. Ann. Appl. Biol. 89: 71–79.

Ehrendorfer, F. 1959. Differentiation-hybridization cycles and polyploidy in *Achillea*. Cold Spring Harb. Symp. Quant. Biol. 24: 141–152.

Ehrendorfer, F. 1962. Beiträge zur Phylogenie der Gattung *Knautia* (Dipsacaceae). I. Cytologische Grundlagen und allgemeine Hinweise. Österr. Bot. Zeit. 109: 276–343.

Ehrendorfer, F. 1965. Dispersal mechanisms, genetic systems, and colonizing abilities in some flowering plant families. Pages 331–351 in H.G. Baker and G.J. Stebbins, eds., The Genetics of Colonizing Species. Academic Press, New York.

Ehrendorfer, F. 1980. Polyploidy and distribution. Pages 45–60 in W.H. Lewis, ed., Polyploidy—Biological Relevance. Plenum, New York.

Einset, J. 1952. Spontaneous polyploidy in cultivated apples. Am. Soc. Hort. Sci. 59: 291–302.

Ekdahl, I.V. 1944. Comparative studies in diploid and tetraploid barley. Ark. Bot. 31: 1–45.

Ekdahl, I.V. 1949. Gigas properties and acreage yield in autotetraploid *Galeopsis pubescens.* Hereditas 35: 397–421.

Ellstrand, N.C., and D.A. Levin. 1980a. Recombination system and population structure in *Oenothera.* Evolution 34: 923–933.

Ellstrand, N.C., and D.A. Levin. 1980b. Evolution of *Oenothera laciniata,* a permanent translocation heterozygote. Syst. Bot. 5: 6–16.

Ellstrand, N.C., and D.A. Levin. 1982. Genotypic diversity in *Oenothera laciniata* (Onagraceae), a permanent translocation heterozygote. Evolution 36: 63–69.

Endrizzi, J.E., E.L. Turcotte, and R.J. Kohel. 1984. Qualitative genetics, cytology, and cytogenetics. Pages 81–129 in R.J. Kohel and C.F. Lewis, eds., Cotton, ASA/CSSA/SSSA Publishers, Madison, Wisc.

Erickson, L.R., N.A. Strauss, and W.D. Beversdorf. 1983. Restriction patterns reveal origins of chloroplast genomes in *Brassica* amphiploids. Theor. Appl. Genet. 65: 201–206.

Eshed, Y., M. Abu-Abied, Y. Saranga, and D. Zamir. 1992. *Lycopersicon esculentum* lines containing small overlapping introgressions from *L. pennelli.* Theor. Appl. Genet. 83: 1027–1034.

Esselman, E.J., and D.J. Crawford. 1997. Molecular and morphological evidence for the origin of *Solidago albopilosa* (Asteraceae), a rare endemic of Kentucky. Syst. Bot. 22: 245–257.

Evans, A.M. 1962. Species hybridization in *Trifolium.* II. Investigating the prefertilization barriers to compatibility. Euphytica 11: 256–262.

Evans, G.M., A. Durant, and H. Rees. 1966. Associated nuclear changes in the induction of flax genotrophs. Nature 212: 697–699.

Evans, W.C. 1989. Trease and Evans Pharmacognosy. English Language Book Society. Bailliere Tindal, London.

Felber, F. 1991. Establishment of a tetraploid cytotype in a diploid population: effect of the relative fitness of cytotypes. J. Evol. Biol. 4: 195–207.

Felber, F., and J.D. Bever. 1997. Effect of triploid fitness on the coexistence of diploids and tetraploids. Biol. J. Linn. Soc. 60: 95–106.

Feldman, M. 1993. Cytogenetic activity and mode of action of the pairing homoeologous (*Ph1*) gene of wheat. Crop Sci. 33: 894–897.

Feldman, M., G. Galili, and A. Levy. 1986. Genetic and evolutionary aspects of allopolyploidy in wheat. Pages 83–100 in C. Barigozzi, ed., The Origin and Domestication of Cultivated Plants. Elsevier, Amsterdam.

Ferguson, D., and T. Sang. 2001. Speciation through homoploid hybridization between allotetraploids in peonies (*Paeonia*). Proc. Natl. Acad. Sci. USA 98: 3915–3919.

Flavell, R.B. 1980. The molecular characterization and organization of plant chromosomal DNA sequences. Annu. Rev. Plant Physiol. 31: 569–596.

Flavell, R.B. 1986. Repetitive DNA and chromosome evolution in plants. Philos. Trans. R. Soc. Lond. B 312: 227–242.

Ford, V.S., and L.D. Gottlieb. 1999. Molecular characterization of *PgiC* in a tetraploid plant and its diploid relatives. Evolution 53: 1060–1067.

Fowler, N.L., and D.A. Levin. 1984. Ecological constraints on the establishment of a novel polyploid in competition with its diploid progenitor. Am. Nat. 124: 703–711.

Frame, J. 1976. The potential of tetraploid red clover and its role in the United Kingdom. J. Br. Grassl. Soc. 31: 139–152.

Franzke, A., and K. Mummenhoff. 1999. Recent hybrid speciation in *Cardamine* (Brassicaceae)—conversion of nuclear ribosomal ITS sequences in *statu nascendi*. Theor. Appl. Genet. 98: 831–834.

Frediani, M., N. Colonna, R. Cremonini, C. De Pace, V. Delre, R. Caccia, and P.G. Cionini. 1994. Redundancy modulation of nuclear DNA sequences in *Dasypyrum villosum*. Theor. Appl. Genet. 88: 167–174.

Frydrych, J. 1965. The study of photosynthetic assimilation of diploid and autotetraploid *Raphanus sativa* L. Genet. Slechteni 2: 25–30.

Fukuda, I. 1967. The biosystematics of *Achlys*. Taxon 16: 308–316.

Fukuda, I. 1984. Chromosome banding and biosystematics. Pages 97–116 in W.F. Grant, ed., Plant Biosystematics. Academic Press, New York.

Gadella, T.W.J. 1987. Sexual tetraploid and apomictic pentaploid populations of *Hieracium pilosella* (Compositae). Plant Syst. Evol. 157: 219–245.

Gadella, W.J., and E. Kliphius. 1968. *Parnassia palustris* in the Netherlands. Acta. Bot. Neerl. 17: 165–172.

Gale, J.S. 1990. Theoretical Population Genetics. Unwin Hyman, London.

Ganal, M.W., N.L.V. Lapitan, and S.D. Tanksley. 1988. A molecular and cytogenetic survey of major repeated DNA sequences in tomato (*Lycopersicon esculentum*). Mol. Gen. Genet. 213: 262–268.

Gajewski, W. 1959. Evolution in the genus *Geum*. Evolution 13: 378–388.

Garbutt, K., and F.A. Bazzaz. 1983. Leaf demography, flower production and biomass of diploid and tetraploid populations of *Phlox drummondii* Hook. on a soil moisture gradient. New Phytol. 93: 129–141.

García-Ramos, G., and M. Kirkpatrick. 1997. Genetic models of adaptation and gene flow in peripheral populations. Evolution 51: 21–28.

Gastony, G.J. 1991. Gene silencing in a polyploid homosporous fern: paleopolyploidy revisited. Proc. Natl. Acad. Sci. USA 88: 1602–1605.

Gatt, M., H. Ding, K. Hammett, and B. Murray. 1998. Polyploidy and evolution in wild and cultivated *Dahlia* species. Ann. Bot. 81: 647–656.

Gaut, B.S, and J.F. Doebley. 1997. DNA sequence evidence for the segmental allotetraploid origin of maize. Proc. Natl. Acad. Sci. USA 94: 6809–6814.

Gaut, B.S., M. Le Thierry d'Ennequin, A.S. Peek, and M.C. Sawkins. 2000. Maize as a model for the evolution of plant nuclear genomes. Proc. Natl. Acad. Sci. USA 97: 7008–7015.

Gauthier, P., R. Lumaret, and A. Be'décarrats. 1998. Genetic variation and gene flow in Alpine diploid and tetraploid populations of *Lotus* (*L. alpinus* (D.C.) Schleicher/ *L. corniculatus* L.) I. Insights from morphological and allozyme markers. Heredity 80: 683–693.

Gebhardt, C., E. Ritter, A. Barone, T. Debner, B. Walkemeier, U. Schachtschabel, H. Kaufmann, R.D. Thompson, M.W. Bonierbale, M.W. Ganal, and S.D. Tanksley. 1991. RFLP maps of potato and their alignment with the homoeologous tomato genome. Theor. Appl. Genet. 83: 49–57.

Geever, R.F., F.R.H. Katterman, and J.E. Endrizzi. 1989. DNA hybridization analyses of a *Gossypium* allotetraploid and two closely related diploid species. Theor. Appl. Genet. 77: 553–559.

Geitler, L. 1937. Cytogenetische Untersuchungen an natürlichen Populationen von *Paris quadrifolia*. Zeitschr. Ind. Abst. Vererbungsl. 73: 182–197.

Geitler, L. 1938. Weitere cytogenetische Untersuchungen an natürlichen Populationen von *Paris quadrifolia*. Zeitschr. Ind. Abst. Vererbungsl. 75: 161–190.

Gerats, A.G.M., H. Huits, E. Vrijlandt, C. Mamrana, E. Souer, and M. Beld. 1990. Molecular characterization of a nonautonomous transposable element (*dtph1*) of petunia. Plant Cell 2: 1121–1128.

Gerassimova, H. 1935. The nature and causes of mutations. II. Transmission of mutations arising in aged seeds: occurrence of "homozygous dislocants" among progeny of plants raised from aged seeds. Cytologia 6: 431–437.

Gerstel, D.U., and J.A. Burns. 1966. Chromosomes of unusual length in hybrids between two species of *Nicotiana*. Chromosomes Today 1: 41–56.

Gillies, C.B., J. Kuspira, and R.N. Bhambhani. 1987. Genetic and cytogenetic analyses of the A genome of *Triticum monococcum*. IV. Synaptonemal complex in autotetraploids. Genome 29: 309–318.

Givins, J.F., and R.L. Phillips. 1976. The nucleolus organizing region of maize (*Zea mays* L.). Chromosoma 57: 103–157.

Goldblatt, P. 1980. Polyploidy in angiosperms. Pages 219–239 in W.H. Lewis, ed., Polyploidy—Biological Relevance. Plenum, New York.

Goldblatt, P., and M. Takei. 1993. Chromosome cytology of the African genus *Lapeirousia* (Iridaceae-Ixioideae). Ann. Missouri Bot. Gard. 80: 961–973.

Goldblatt, P., and M. Takei. 1997. Chromosome cytology of Iridaceae—patterns of variation, determination of ancestral base numbers, and modes of karyotype change. Ann. Missouri Bot. Gard. 84: 285–304.

Gómez, M.I., M.N. Islam-Faridi, M.S. Zwick, D.G. Czeschin Jr., G.E. Hart, R.A. Wing, D.M. Stelly, and H.J, Price. 1998. Tetraploid nature of *Sorghum bicolor* (L.) Moench. J. Hered. 89: 188–190.

Goodwillie, C. 1999. Multiple origins of self-compatibility in *Linanthus* section *Leptosiphon* (Polemoniaceae): phylogenetic evidence from internal-transcribed-spacer sequence data. Evolution 53: 1387–1395.

Goral, S., T. Hulewicz, and E. Polakowska. 1964. Winter hardiness and chemical composition of diploid and polyploid red alsike, and white clover during the winter season. Genet. Polon. 5: 289–307.

Görg, R., K. Hollricher, and P. Schulze-Lefert. 1993. Functional analysis and RFLP-mediated mapping of the *Mlg* resistant locus in barley. Plant J. 3: 857–866.

Gottlieb, L.D. 1974. The genetic confirmation of the origin of *Clarkia lingulata*. Evolution 28: 244–250.

Gottlieb, L.D., and V.S. Ford. 1996. Phylogenetic relationships among the sections of *Clarkia* (Onagraceae) inferred from the nucleotide sequences of *PgiC*. Syst. Bot. 21: 45–62.

Gottlieb, L.D., and V.S. Ford. 1997. A recently silenced, duplicate *PgiC* locus in *Clarkia*. Mol. Biol. Evol. 14: 125–132.

Gottschalk, W. 1951. Untersuchungen am Pacytän normaler und rontgenbestrahlter Pollenmutterzellern von *Solanum lycopersicum*. Chromosoma 4: 298–341.

Gottschalk, W. 1976. Die Bedeutung der Polyploidie fur die Evolution der Pflanzen. Gustav Fisher Verlag, Stuttgart.

Gould, F.W. 1966. Chromosome numbers of some Mexican grasses. Can. J. Bot. 44: 1683–1696.

Gould, F.W., and Z.J. Kapadia. 1962. Biosystematic studies in the *Bouteloua curtipendula* complex. I. The aneuploid rhizomatous *B. curtipendula* of Texas. Am. J. Bot. 49: 887–892.

Gould, F.W., and Z.J. Kapadia. 1964. Biosystematic studies in the *Bouteloua curtipendula* complex. II. Taxonomy. Brittonia 16: 182–207.

Grant, D., P. Cregan, and R.C. Shoemaker. 2000. Genome organization in dicots: genome duplication in *Arabidopsis* and synteny between soybean and *Arabidopsis*. Proc. Natl. Acad. Sci. USA 97: 4168–4173.

Grant, V. 1952. Cytogenetics of the hybrid *Gilia millefoliata* × *achilleaefolia*. I. Variations in meiosis and polyploidy rates as affected by nutritional and genetic conditions. Chromosoma 5: 372–390.

Grant, V. 1954. Genetic and taxonomic studies in *Gilia*. IV. *Gilia achilleaefolia*. Aliso 3: 1–18.

Grant, V. 1956. The genetic structure of races and species in *Gilia*. Adv. Genet. 6: 55–87.

Grant, V. 1963. The Origin of Adaptations. Columbia University Press, New York.

Grant, V. 1966a. Selection for vigor and fertility in the progeny of a highly sterile species hybrid in *Gilia*. Genetics 53: 757–775.

Grant, V. 1966b. The origin of a new species of *Gilia* in a hybridization experiment. Genetics 54: 1189–1199.

Grant, V. 1975. Genetics of Flowering Plants. Columbia University Press, New York.

Grant, V. 1981. Plant Speciation, 2nd ed. Columbia University Press, New York.

Grattapaglia, D., and H.D. Bradshaw. 1994. Nuclear DNA content of commercially important *Eucalyptus* species and their hybrids. Can. J. Forest Res. 24: 1074–1078.

Grau, J. 1964. Die Zytotaxonomie der *Myosotis alpestris* und der *Myosotis sylvaticus*-Gruppe in Europa. Österr. Bot. Zeit. 111: 561–617.

Griesbach, R.J., and. K.K. Kamo. 1996. The effect of induced polyploidy on the flavonols of *Petunia* "Mitchell." Phytochem. 42: 361–363.

Grif, V.G. 2000. Some aspects of plant karyology and karyosystematics. Int. Rev. Cytol. 196: 131–175.

Grime, J.P., and M.A. Mowforth. 1982. Variation in genome size—an ecological interpretation. Nature 299: 151–153.

Grime, J.P., J.M.L. Shacklock, and S.R. Brand. 1985. Nuclear DNA content, shoot phenology, and species coexistence in a limestone grassland community. New Phytol. 100: 435–445.

Gunthardt, H., L. Smith, M.E. Haferkamp, and R.H. Nolan. 1953. Studies on aged seeds. II. Relation of age of seeds to cytogenic effects. Agron. J. 45: 438–441.

Guo, M., D. Davis, and J.A. Birchler. 1996. Dosage effects on gene expression in a maize ploidy series. Genetics 142: 1349–1355.

Gupta, P.O. 1981. Consequences of artificial and natural chromosome doubling on DNA, RNA, and protein contents in *Cochlearia* (Brassiceae). Plant Syst. Evol. 138: 23–27.

Gustafsson, M. 1972. Distribution and effects of pericentric inversions in populations of *Atriplex longipes*. Hereditas 71: 173–194.

Gustafsson, M. 1974. Evolutionary trends in the *Atriplex triangularis* group of Scandinavia. III. The effects of population size and introgression on chromosomal differentiation. Bot. Not. 127: 125–148.

Hagberg, A., and S. Ellerström. 1959. The competition between diploid, tetraploid and aneuploid rye. Hereditas 45: 369–416.

Hagerup, O. 1927. *Empetrum hermaphroditum* (Lge.) Hagerup. A new tetraploid, bisexual species. Dansk. Bot. Arkiv. 5: 1–17.

Hagerup, O. 1932. Über Polyploidie in Beziehung zu Klima, Okologie und Phylogenie. Hereditas 16: 19–50.

Hagerup, O. 1933. Studies on polyploid ecotypes in *Vaccineum uliginosum* L. Hereditas 18: 122–128.

Hair, J.B. 1966. Biosystematics of the New Zealand flora 1945–1964. N.Z. J. Bot. 4: 559–595.

Haldane, J.B.S. 1930. Theoretical genetics of autotetraploids. J. Genet. 22: 359–372.

Hall, O. 1972. Oxygen requirement of root meristems in diploid and tetraploid rye. Hereditas 70: 69–74.

Hanelt, P. 1966. Polyploidie-Frequenz und geographische Verbretung bei höheren Pflanzen. Biol. Rundschau 4: 183–196.

Hanneman, R.K., Jr. 1994. Assignment of Endosperm Balance Numbers to the tuber-bearing solanums and their close non-tuber-bearing relatives. Euphytica 74: 19–25.

Hanson, R.E., X.-P. Zhao, M.N. Islam-Faridi, A.H. Paterson, M.S. Zwick, C.F. Crane, T.D. McKnight, D.M. Stelly, and H.J. Price. 1998. Evolution of interspersed repetitive elements in *Gossypium* (Malvaceae). Am. J. Bot. 85: 1364–1368.

Hanson, W.D. 1959a. Early generation analysis of lengths of chromosome segments around a locus held heterozygous with backcrossing or selfing. Genetics 44: 833–837.

Hanson, W.D. 1959b. The breakup of initial linkage blocks under selected mating systems. Genetics 44: 857–868.

Harberd, N.P., R.B. Flavell, and R.D. Thompson. 1987. Identification of a transposon-like insertion in a *Glu-1* allele of wheat. Mol. Gen. Genet. 209: 326–332.

Harborne, J.B., C.A. Williams, and D.M. Smith. 1973. Species-specific kaempferol derivatives in ferns of the Appalachian *Asplenium* complex. Biochem. Syst. Ecol. 1: 51–54.

Hardy, O.J., S. Vanderhoeven, M. De Loose, and P. Meerts. 2000. Ecological, morphological, and allozymic differentiation between diploid and tetraploid knapweeds (*Centaurea jacea*) from a contact zone in the Belgian Ardennes. New Phytol. 146: 281–290.

Harlan, J.R., and J.M. deWet. 1975. On Ö. Winge and a prayer: the origins of polyploidy. Bot. Rev. 41: 361–390.

Hart, G.E. 1983. Hexaploid wheat. Pages 35–56 in S. Tanksley and T.J. Orton, eds., Isozymes in Plant Genetics and Breeding. Part B. Liss, New York.

Hashemi, A., A. Estilai, and J. Waines. 1989. Cytogenetics and reproductive behavior of induced and natural tetraploid guayule (*Parthenium argentatum* Gray). Genome 32: 1100–1104.

Haskell, G. 1968. Biochemical differences between spontaneous and colchicine-induced autotetraploids. Heredity 23: 139–141.

Hauber, D.P. 1986. Autotetraploidy in *Haplopappus spinulosus* hybrids: evidence from natural and synthetic tetraploids. Am. J. Bot. 73: 1595–1606.

Hawkes, J.G. 1990. The Potato: Evolution, Biodiversity, and Genetic Resources. Belhaven Press, London.

Hawkes, J.G., and M.T. Jackson. 1992. Taxonomic and evolutionary implications of the endosperm balance number hypothesis in potatoes. Theor. Appl. Genet. 84: 180–185.

Hazarika, M.H., and H. Rees. 1967. Genotypic control of chromosome behaviour in rye. X. Chromosome pairing and fertility in autotetraploids. Heredity 22: 317–332.

Heckard, L.R. 1960. Taxonomic studies in the *Phacelia magellanica* polyploid complex with special reference to the California members. Univ. Calif. Publ. Bot. 32: 1–126.

Hedberg, I. 1969. Cytotaxonomic studies on *Anthoxanthum odoratum* L. s. lat. III. Investigations of Swiss and Austrian population samples. Sv. Bot. Tidskr. 63: 233–250.

Hedrick, P.W. 1981. The establishment of chromosomal variants. Evolution 35: 322–332.

Hedrick, P.W., and D.A. Levin. 1984. Kin-founding and the fixation of chromosomal variants. Am. Nat. 124: 789–797.

Hecht, A. 1950. Cytogenetic studies of *Oenothera*, subgenus *Raimannia*. Indiana Univ. Publ. Sci. Serv. 16: 225–304.

Heiser, C.B. 1956. Biosystematics of *Helianthus debilis*. Madroño 13: 145–176.

Heiser, C.B., W.C. Martin, and D.M. Smith. 1962. Species crosses in *Helianthus*. I. Diploid species. Brittonia 14: 137–147.

Helentjaris, T., D. Weber, and S. Wright. 1988. Identification of genomic locations of duplicate nucleotide sequences in maize by analysis of restriction length polymorphisms. Genetics 118: 353–363.

Hemleben, V. 1990. Molekularbiology der Pflanzen. Gustav Fischer Verlag, Stuttgart.

Heneen, W.K. 1972. Chromosomal polymorphism in isolated populations of *Elymus* in the Aegean. I. *Elymus striatus* sp. nov. Bot. Not. 125: 419–429.

Heneen, W.K., and H. Runemark. 1962. Chromosomal polymorphism and morphological diversity in *Elymus rechingeri*. Hereditas 48: 545–564.

Hewitt, G.M. 1988. Hybrid zones—natural laboratories for evolutionary studies. Trends Ecol. Evol. 3: 158–167.

Hewitt, G.M. 1993. Postglacial distribution and species substructure: lessons from pollen, insects, and hybrid zones. Pages 98–123 in D.R. Lees and D. Edwards, eds., Evolutionary Patterns and Processes. The Linnean Society of London.

Heywood, J.S. 1986. The effect of plant size variation on genetic drift in populations of annuals. Am. Nat. 127: 851–861.

Heywood, J.S., and D.A. Levin. 1984. Allozyme variation in *Gaillardia pulchella* and *G. amblyodon* (Compositae): relation to morphological and chromosomal variation and to geographical isolation. Syst. Bot. 9: 448–457.

Hiesey, W.M., M.A. Nobs, and O. Björkman. 1971. Experimental studies on the nature of species. V. Biosystematics, genetics and physiological ecology of the *Eurythranthe* section *Mimulus*. Publ. 628. Carnegie Institution, Washington, DC.

Hill, R.R. 1971. Selection in autotetraploids. Theor. Appl. Genet. 41: 181–186.

Hilpert, G. 1957. Effect of selection for meiotic behaviour in autotetraploid rye. Hereditas 43: 318–322.

Hrishi, N. 1969. Further data on structural heterozygosity in a strain of *Secale kuprijanovii*. Hereditas 62: 339–347.

Humphreys, M.W., H.M. Thomas, W.G. Morgan, M.R. Meredith, J.A. Harper, H. Thomas, Z. Zwierzkowski, and M. Ghesquire. 1995. Discriminating the ancestral progenitors of hexaploid *Festuca arundinacea* using genomic *in situ* hybridization. Heredity 75: 171–174.

Hutchinson, J., H. Rees, and A.G. Seal. 1979. An assay of the activity of supplementary DNA in *Lolium*. Heredity 43: 411–421.

Hodgson, J.G. 1987. Why do so few plant species exploit productive habitats? An investigation into cytology, plant strategies and abundance within a local flora. Functional Ecol. 1: 243–250.

Holm, P.B., and X. Wang. 1988. The effect of chromosome 5B on synapsis and chiasma formation in wheat. Carlsberg Res. Comm. 53: 191–208.

Holsinger, K.E., and N.C. Ellstrand. 1984. The evolution and ecology of permanent translocation heterozygotes. Am. Nat. 124: 48–71.

Holsinger, K.E., and M.W. Feldman. 1981. A single locus model of selection in permanent translocation heterozygotes. Theor. Pop. Biol. 20: 218–240.

Husband, B.C. 2000. Constraints on polyploid evolution: a test of the minority cytotype exclusion principle. Proc. R. Soc. Lond. B. 267: 217–223.

Husband, B.C., and D.W. Schemske. 1997. The effect of inbreeding in diploid and tetraploid populations of *Epilobium angustifolium* (Onagraceae): implications for the genetic basis of inbreeding depression. Evolution 51: 737–746.

Husband, B.C., and D.W. Schemske. 1998. Cytotype distribution at a diploid-tetraploid contact zone in *Chamerion* (*Epilobium*) *angustifolium* (Onagraceae). Am. J. Bot. 85: 1688–1694.

Hutchinson, J., H. Rees, and A.G. Seal. 1979. An assay of the activity of supplementary DNA in *Lolium*. Heredity 43: 411–421.

Hutton, E.M., and J.W. Peak. 1954. The effect of autopolyploidy in five varieties of subterranean clover (*Trifolium subterraneam* L.). Aust. J. Agric. Res. 5: 356–364.

Hymowitz, T., R.J. Singh, and K.P. Kollipara. 1998. Genomes of *Glycine*. Plant Breed. Rev. 16: 289–317.

Ingram, R., and H.J. Noltie. 1989. Early adjustment of patterns of metaphase association in the evolution of a polyploid species. Genetica 78: 21–24.

Iwanaga, M., Y. Mukai, I. Panayotov, and K. Tsunewaki. 1978. Genetic diversity of the cytoplasm in *Triticum* and *Aegilops*. VII. Cytoplasmic effects on respiratory and photosynthetic rates. Jpn J. Genet. 53: 387–396.

Iyengar, N.K. 1944. Cytological investigations on auto- and allotetraploid Asiatic cottons. Indian J. Agric. Sci. 14: 30–40.

Jackson, R.C. 1962. Interspecific hybridization in *Haplopappus* and its bearing on chromosome evolution in the *Blepharodon* section. Am. J. Bot. 49: 119–132.

Jackson, R.C. 1965. A cytogenetic study of a three-paired race of *Haplopappus gracilis*. Am. J. Bot. 52: 946–953.

Jackson, R.C. 1973. Chromosomal evolution in *Haplopappus gracilis:* a centric transposition race. Evolution 27: 243–256.

Jackson, R.C. 1985. Genomic differentiation and its effect on gene flow. Syst. Bot. 10: 391–404.

Jackson, J.C., and J. Casey. 1982. Cytogenetic analyses of autopolyploids: models and methods for triploids to octoploids. Am. J. Bot. 69: 489–503.

Jackson, R.C., and D.P. Hauber. 1982. Autotriploid and autotetraploid cytogenetic analysis: correction for proposed binomial models. Am. J. Bot. 69: 644–646.

Jackson, R.C, and J.W. Jackson. 1996. Gene segregation in autotetraploids: prediction from meiotic configurations. Am. J. Bot. 83: 673–678.

Jackson, R.C., and B. Murray. 1983. Colchicine-induced quadrivalent formation in *Helianthus:* evidence of ancient polyploidy. Theor. Appl. Genet. 64: 219–222.

James, S.H. 1965. Complex hybridity in *Isotoma petraea*. I. The occurrence of interchange heterozygotes, autogamy, and a balanced lethal system. Heredity 20: 341–353.

James, S.H., J.F. Sampson, and J. Playford. 1990. Complex hybridity in *Isotoma petraea*. VII. Assembly of the genetic system in the O_6 Pidgeon Rock population. Heredity 64: 289–295.

James, S.H., A.P. Wylie, M.S. Johnson, S.A. Carstairs, and G.A. Simpson. 1983. Complex hybridity in *Isotoma petraea*. V. Allozyme variation and the pursuit of hybridity. Heredity 51: 653–663.

Jan, C.C. 1992. Cytoplasmic-nuclear gene interaction for plant vigor in *Helianthus* species. Crop. Sci. 32: 320–323.

Janaki Ammal, E.K., and S.N. Sobti. 1962. The origin of Jammu mint. Curr. Sci. 31: 387–388.

Jellen, E.N., B.S. Gill, and T.S. Cox. 1994. Genomic *in situ* hybridization differentiates between A/D- and C-genome chromatin and detects intergenomic translocations in polyploid oat species (genus *Avena*). Genome 37: 613–618.

Jenkins, G. 1985. Synaptonemal complex formation in hybrids of *Lolium temulentum* × *Lolium perenne* (L.). Chromosoma 92: 81–88.

Jenkins, G., and G. Jimenez. 1995. Genetic control of synapsis and recombination in *Lolium* amphidiploids. Chromosoma 104: 164–168.

Jiang, C., R.J. Wright, K.J. El-Zik, and A.H. Paterson. 1998. Polyploid formation created unique avenues for response to selection in *Gossypium*. Proc. Natl. Acad. Sci. USA 95: 4419–4424.

Jiang, J., and B.S. Gill. 1994. Different species-specific chromosome translocations in *Triticum timopheevii* and *T. turgidum* support the diphyletic origin of polyploid wheats. Chromosome Res. 2: 59–64.

Johnsson, H. 1944. Meiotic aberrations and sterility in *Alopecurus myosuroides*. Huds. Hereditas 30: 469–566.

Johnsson, H. 1945. Chromosome numbers of the progeny from the cross triploid × tetraploid *Populus tremula*. Hereditas 31: 500–501.

Johnston, S.A., T.P.N. den Nijs, S.J. Peloquin, and R.E. Hanneman 1 1980. The significance of genetic balance to endosperm development in interspecific crosses. Theor. Appl. Genet. 57: 5–9.

Johnston, S.A., and R.E. Hanneman Jr. 1980. Support of the Endosperm Balance Number hypothesis utilizing some tuber-bearing *Solanum* species. Am. Potato J. 57: 7–14.

Johnston, S.A., and R.E. Hanneman Jr. 1982. Manipulations of Endosperm Balance Number overcome crossing barriers between diploid *Solanum* species. Science 217: 446–448.

Johnston, S.A., and R.E. Hanneman Jr. 1996. Genetic control of Endosperm Balance Number (EBN) in the *Solanaceae* based on trisomic and mutation analysis. Genome 39: 314–321.

Jones, G.H. 1969. Further correlations between chiasmata and U-type exchanges in rye meiosis. Chromosoma 26: 105–118.

Jones, G.H., and J.E. Vincent. 1994. Meiosis in autotetraploid *Crepis capillaris*. II. Autotetraploids. Genome 37: 497–505.

Jones, K. 1998. Robertsonian fusion and centric fission in karyotype evolution of higher plants. Bot. Rev. 64: 273–289.

Jones, R.N., and L.M. Brown. 1976. Chromosome evolution and DNA variation in *Crepis*. Heredity 36: 91–104.

Jones, K., D. Papes, and D.R. Hunt. 1975. Contributions to the cytotaxonomy of the Commelinaceae. II. Further observations on *Gibasis geniculata* and its allies. Bot. J. Linn. Soc. 71: 145–166.

Jongedijk, E., M.S. Ramanna, Z. Sawor, and J.G.T. Hermsen. 1991. Formation of first division restitution (FDR) 2*n*-megaspores through pseudohomotypic division in ds-1 (desynapsis) mutants of diploid potato: routine production of tetraploid progeny from 2*x*FDR × 2*x*FDR crosses. Theor. Appl. Genet. 82: 645–656.

Jorgensen, C.A. 1928. The experimental formation of hyperploid plants in the genus *Solanum*. J. Genet. 19: 133–211.

Joseph, M.C., D.D. Randall, and C. Nelson. 1981. Photosynthesis in polyploid tall fescue. II. Photosynthesis and ribulose-1,5-bisphosphate carboxylase of polyploid tall fescue. Plant Physiol. 68: 894–898.

Kalendar, R., J. Tanskanen, S. Immonen, and A.H. Schulman. 2000. Genome evolution of wild barley (*Hordeum spontaneum*) by BARE-1 retrotranspon dynamics in response to sharp microclimatic divergence. Proc. Natl. Acad. Sci. USA 97: 6603–6607.

Kalton, R.R., A.G. Smit, and R.C. Leffel. 1952. Parent-inbred progeny relationships of selected orchardgrass clones. Agron. J. 44: 481–486.

Katsiotis, A., R.E. Hanneman Jr., and R.A. Fosberg. 1995. Endosperm Balance Number and the polar-nuclei activation hypotheses for endosperm development in interspecific crosses of Solanaceae and Gramineae, respectively. Theor. Appl. Genet. 91: 848–855.

Kaul, M.L.H., and T.G.K. Murthy. 1985. Mutant genes affecting higher plant meiosis. Theor. Appl. Genet. 70: 449–466.

Kay, Q.O.N. 1969. The origin and distribution of diploid and tetraploid *Tripleuronspermum inodortum* (L.) Schultz Bip. Watsonia 7: 130–141.

Kellogg, E.A., R. Appells, and R.J. Mason-Gomer. 1996. When genes tell different stories: the diploid genera of Triticeae (Gramineae). Syst. Bot 21: 1–17.

Kenton, A. A.S. Parokonny, Y.Y. Gleba, and M.D. Bennett. 1993. Characterization of the *Nicotiana tabacum* L. genome by molecular cytogenetics. Mol. Gen. Genet. 240: 159–169.

Kihara, H., and T. Ono. 1926. Chromosomenzahlii und systematische Gruppierung der Rumex-Arten. Z. Zellforsch. Mikr. Anat. 4: 475–481.

Khawaja, J.I.T., J. Sybenga, and J.R. Ellis. 1997. Chromosome pairing and chiasmata formation in autopolyploids of different *Lathyrus* species. Genome 40: 937–944.

Khoshoo, T.N. 1962. Cytogenetical evolution in gymnosperms-karyotype. Pages 119–135 in: Proceedings of Summer School, Darjeeling. Government of India, New Delhi.

Khoshoo, T.N., and V.B. Sharma. 1959. Biosystematics of *Sisymbrium irio* complex. Reciprocal pollinations and seed failure. Caryologia 12: 71–97.

Kim, J.-K, and R.K. Jansen. 1994. Comparison of phylogenetic hypotheses among different data sets in dwarf dandelions (*Krigia*, Asteraceae): additional information from internal transcribed spacer sequences of nuclear ribosomal DNA. Plant Syst. Evol. 190: 157–185.

Kirkpatrick, M., and N.H. Barton. 1997. Evolution of a species' range. Am. Nat. 150: 1–23.

Kollipara, K.P., R.J. Singh, and T. Hymowitz. 1994. Genomic diversity and multiple origins of tetraploid ($2n=78,80$) *Glycine tomentella*. Genome 37: 448–459.

Konishi, N., K. Watanabe, and K. Kosunge. 2000. Molecular systematics of Australian *Podolepis* (Asteraceae: Gnaphalieae): evidence from DNA sequences of the nuclear ITS region and the chloroplast *matK* gene. Austral. Syst. Bot. 13: 709–727.

Kostoff, D., and J. Kendall. 1929. Irregular meiosis in *Lycium halimifolium* Mill. produced by gall mites (Eriophyes). J. Genet. 21: 113–115.

Kostoff, D., and J. Kendall. 1934. Studies on polyploid plants. III. Cytogenetics of tetraploid tomatoes. Gartenbauwissenschaft 9: 1–20.

Kreike, C.M., and W.J. Stiekema. 1997. Reduced recombination and distorted segregation in a *Solanum tuberosum* ($2x$) × *S. spegazzinii* ($2x$) hybrid. Genome 40: 180–187.

Ku, H.-M., T. Vision, J. Liu, and S.D. Tansley. 2000. Comparing sequenced segments of the tomato and *Arabidopsis* genomes: large-scale duplication followed by selective gene loss creates a network of synteny. Proc. Natl. Acad. Sci. USA 97: 9121–9126.

Kubis, K., T. Schmidt, and J.S. Heslop-Harrison. 1998. Repetitive DNA elements as a major component of plant genomes. Ann. Bot. 82 (Suppl. A): 45–55.

Kulkarni, R.N., and N.S. Ravindra. 1988. Resistance to *Pythium alphanidertum* in diploids and induced autotetraploids of *Catharanthus roseus*. Planta Med. 54: 356–359.

Kunze, R., H. Saedler, and W.-E. Lönnig. 1997. Plant transposable elements. Adv. Bot. Res. 27: 332–470.

Kurabayashi, M., H. Lewis, and P.H. Raven. 1962. A comparative study of meiosis in the Onagraceae. Am. J. Bot. 49: 1003–1026.

Kurita, M., and Y. Kuroki. 1964. Polyploidy and distribution of *Allium gray[i]*. Mem. Ehime Univ. Sect. 2 Ser. B. 5: 37–45.

Kyhos, D.W. 1965. The independent aneuploid origin of two species of *Chaenactis* (Compositae) from a common ancestor. Evolution 19: 26–43.

Kyhos, D.W., and G.D. Carr. 1994. Chromosome stability and lability in plants. Evol. Theory 10: 227–248.

Kyhos, D.W., G.D. Carr, and B.G. Baldwin. 1990. Biodiversity and cytogenetics of the tarweeds (Asteraceae: Heliantheae-Madiinae). Ann. Missouri Bot. Gard. 77: 84–95.

Lagercrantz, U. 1998. Comparative mapping between *Arabidopsis thaliana* and *Brassica nigra* indicates that *Brassica* genomes have evolved through extensive genome replication accompanied by chromosome fusions and frequent rearrangements. Genetics 150: 1217–1228.

Lagercrantz, U., and D.J. Lydiate. 1996. Comparative genomic mapping in *Brassica*. Genetics 144: 1903–1910.

Lande, R. 1979. Effective deme size during long-term evolution estimated from rates of chromosomal rearrangement. Evolution 33: 234–251.

Lande, R. 1985. The fixation of chromosomal rearrangements in a subdivided population with local extinction and colonization. Heredity 54: 323–332.

Lande, R., and D.W. Schemske. 1985. The evolution of self-fertilization and inbreeding depression in plants. I. Genetic models. Evolution 39: 24–40.

Lapitan, N. 1992. Organization and evolution of higher plant nuclear genomes. Genome 35: 171–181.

Lapitan, N.L.V., R.G. Sears, and B.S. Gill. 1988. Amplification of a repeated DNA sequence in wheat × rye hybrids generated from tissue culture. Theor. Appl. Genet. 75: 381–388.

Laushman, R.H., A. Schnabel, and J.L. Hamrick. 1996. Electrophoretic evidence for tetrasomic inheritance in the dioecious tree *Maclura pomifera* (Raf.) Schneid. J. Hered. 87: 469–473.

Lavania, U.C. 1986. Genetic improvement of the Egyptian henbane, *Hyocyamus muticus* L. through induced tetraploidy. Theor. Appl. Genet. 73: 292–298.

Lavania, U.C. 1988. Enhanced productivity of the essential oil in the artificial autopolyploid of vetiver (*Vetiveria zizanoides* L. Nash). Euphytica 38: 271–276.

Lavania, U.C., and S. Srivastava. 1991. Enhanced productivity of tropane alkaloids and fertility in artificial autotetraploids of *Hyocyamus niger* L. Euphytica 52: 73–77.

Lawrence, W.J.C. 1958. Studies on *Streptocarpus* Lindl. V. Speciation and gene systems. Heredity 12: 333–356.

Leitch, I.J., M.W. Chase, and M.D. Bennett. 1998. Phylogenetic analysis of DNA C-values provides evidence for a small ancestral genome size in flowering plants. Ann. Bot. 82: 85–94.

Leitch, I.J., and M.D. Bennett. 1997. Polyploidy in angiosperms. Trends Plant Sci. 2: 470–476.

Levan, A. 1932. Cytological studies in *Allium*. I. Chromosome morphological contributions. Hereditas 16: 257–264.

Levan, A. 1935. Cytological studies in *Allium*. VI. Chromosome morphology of some diploid species of *Allium*. Hereditas 20: 289–330.

Levan, A. 1945. Aktvelle Probleme der Polyploidiezüchtung. Arch. Julius Klaus-Stiftg. Erg. Bd. 20: 142–152.

Levin, D.A. 1975a. Minority cytotype exclusion in local plant populations. Taxon 24: 35–43.

Levin, D.A. 1975b. Genetic correlates of translocation heterozygosity in plants. Bioscience 25: 724–728.

Levin, D.A. 1975c. Genic heterozygosity and protein polymorphism among local populations of *Oenothera biennis*. Genetics 79: 477–491.

Levin, D.A. 1978. Some genetic consequences of being a plant. Pages 189–212 in P.F. Brussard, ed., Ecological Genetics: The Interface. Springer-Verlag, New York.

Levin, D.A. 1983. Polyploidy and novelty in flowering plants. Am. Nat. 122: 1–25.

Levin, D.A. 1990. The seed bank as a source of genetic novelty in plants. Am. Nat. 135: 563–572.

Levin, D.A. 2000. The Origin, Expansion, and Demise of Plant Species. Oxford University Press, New York.

Levin, D.A. 2001. The recurrent origin of plant races and species. Syst. Bot. 26: 197–204.

Levin, D.A., and S.W. Funderberg. 1979. Genome size in angiosperms: temperate versus tropical species. Am. Nat. 114: 784–795.

Levin, D.A., and H.W. Kerster. 1974. Gene flow in seed plants. Evol. Biol. 7: 139–220.

Levin, D.A., A.M. Torres, and M. Levy. 1978. Alcohol dehydrogenase activity in diploid and autotetraploid *Phlox*. Biochem. Genet. 17: 35–42.

Levin, D.A., and A.C. Wilson. 1976. Rates of evolution in seed plants: net increase in diversity of chromosome numbers and species numbers through time. Proc. Natl. Acad. Sci. USA 73: 2086–2090.

Levy, M. 1976. Altered glycoflavone expression in induced autotetraploids of *Phlox drummondii*. Biochem. Syst. Evol. 4: 249–259.

Levy, M., and D.A. Levin. 1975. Genic heterozygosity and variation in permanent translocation heterozygotes of the *Oenothera biennis* complex. Genetics 79: 493–512.

Levy, M., and P.L. Winternheimer. 1977. Allozyme linkage disequilibria among chromosome complexes in the permanent translocation heterozygote *Oenothera biennis*. Evolution 31: 465–476.

Lewis, D. 1949. Incompatibility in flowering plants. Biol. Rev. 24: 472–496.

Lewis, H. 1951. The origin of supernumerary chromosomes in natural populations of *Clarkia elegans*. Evolution 5: 142–157.

Lewis, H. 1953. Chromosome phylogeny and habitat preference of *Clarkia*. Evolution 7: 102–109.

Lewis, H., and P.H. Raven. 1958. Rapid evolution in *Clarkia*. Evolution 12: 319–336.

Lewis, H., and M.E. Lewis. 1955. The genus *Clarkia*. Univ. Calif. Publ. Bot. 20: 241–392.

Lewis, H., and M.R. Roberts. 1956. The origin of *Clarkia lingulata*. Evolution 10: 126–138.

Lewis, H., and J. Szweykowski. 1964. The genus *Gayophytum* (Onagraceae). Brittonia 16: 343–392.

Lewis, K.R., and B. John. 1959. Breakdown and restoration of chromosome stability following inbreeding in a locust. Chromosoma 10: 589–618.

Lewis, K.R., and B. John. 1963. Chromosome Marker. Little Brown, Boston.

Lewis, W.H. 1967. Cytocatalytic evolution in plants. Bot. Rev. 33: 105–115.

Lewis, W.H. 1970. Chromosomal drift, a new phenomenon in plants. Science 168: 1115–1116.

Lewis, W.H. 1976. Temporal adaptation correlated with polyploidy in *Claytonia virginica*. Syst. Bot. 1: 340–347.

Lewis, W.H. 1980. Polyploidy in species populations. Pages 103–144 in W.H. Lewis, ed., Polyploidy—Biological Relevance. Plenum, New York.

Lewis, W.H., and E.E. Terrell. 1962. Chromosomal races in eastern North American species of *Hedyotis* (*Houstonia*). Rhodora 64: 313–323.

Li, W.-H., and D. Grauer. 1991. Fundamentals of Molecular Evolution. Sinauer Associates, Sunderland, Mass.

Li, W.-L., G.P. Berlyn, and P.M.S. Ashton. 1996. Polyploids and their structural and physiological character relative to water deficit in *Betula papyrifera* (Betulaceae). Am. J. Bot. 83: 15–20.

Liharska, T., M. Koornneef, M. van Wordragen, A. van Kammen, and P. Zabel. 1996. Tomato chromosome 6: effect of alien chromosomal segments on recombinant frequencies. Genome 39: 485–491.

Lin, Y.B. 1984. Ploidy barrier to endosperm development in maize. Genetics 107: 103–115.

Liu, B., H.M. Piao, F.S. Zhao, J.H. Zhao, and R. Zhao. 1999. Production and molecular characterization of rice lines with introgressed traits from a wild species *Zizania latifolia*. J. Genet. Breed. 53: 279–284.

Liu, B., J.M. Vega, and M. Feldman. 1998a. Rapid genomic changes in newly synthesized amphiploids of *Triticum* and *Aegilops* II. Changes in low-copy coding DNA sequences. Genome 41: 535–542.

Liu, B., J.M. Vega, G. Segal, S. Abbo, M. Rodova, and M. Feldman. 1998b. Rapid genomic changes in newly synthesized amphiploids of *Triticum* and *Aegilops*. I. Changes in low-copy noncoding DNA sequences. Genome 41: 272–277.

Liu, B., and J.F Wendel. 2000. Retrotransposon activation followed by rapid repression in introgressed rice plants. Genome 43: 874–880.

Livingstone, K.D., V.K. Lackney, J.R. Blauth, R. van Wijk, and M.K. Jahn. 1999. Genome mapping in *Capsicum* and the evolution of genome structure in the Solanaceae. Genetics 152: 1183–1202.

Loidl, J. 1991. Coming to grips with a complex matter. A multidisciplinary approach to the synaptonemal complex. Chromosome 100: 289–292.

Loidl, J. 1994. Cytological aspects of meiotic recombination. Experentia 50: 285–294.

Löve, A., and D. Löve. 1957. Arctic polyploidy. Proc. Genet. Soc. Can. 2: 23–27.

Lumaret, R. 1985. Phenotypic and genetic variation within and between populations of the polyploid complex, *Dactylis glomerata* L. Pages 343–354 in J. Haeck, and J.W. Woldendrop, eds., Structure and Functioning of Plant Populations 2. Elsevier/ North Holland, Amsterdam.

Lumaret, R.L. 1988. Cytology, genetics, and evolution in the genus *Dactylis*. Crit. Rev. Plant. Sci. 7: 55–91.

Lumaret, R., J.-L. Guillerm, J. Delay, A.A.L. Loutfi, J. Izco, and M. Jay. 1987. Polyploidy and habitat differentiation in *Dactylis glomerata* L. from Galicia (Spain). Oecologia 73: 436–446.

Lundquist, A. 1966. Heterosis and inbreeding depression in autotetraploid rye. Hereditas 56: 317–366.

Maceira, N.O., A.A. De Hann, R. Lumaret, M. Billion, and J. Delay. 1992. Production of 2n gametes in diploid subspecies of *Dactylis glomerata* L. I. Occurrence and frequency of diploid pollen. Ann. Bot. 69: 335–343.

Maceira, N.O., P. Jacquard, and R. Lumaret. 1993. Competition between diploid and derivative autotetraploid *Dactylis glomerata* L. from Galicia. Implications for the establishment of novel polyploid populations. New Phytol. 124: 321–328.

Macgillivray, C.W., and J.P. Grime. 1995. Genome size predicts frost resistance in British herbaceous plants: implications for rates of vegetation response to global warming. Funct. Ecol. 9: 320–325.

Mackay, I.J. 1980. Population genetics of *Papaver dubium*. Ph.D. thesis, University of Birmingham.

Madahar, C., and M. Heimburger. 1969. Meiotic studies in *Anemone multifida, A. tetonensis,* and their hybrids. Can. J. Bot. 47: 1973–1983.

Maestra, B., and T. Naranjo. 1998. Homoeologous relationships of *Aegilops speltoides* chromosomes to bread wheat. Theor. Appl. Genet. 97: 181–186.

Magoon, M.L., S. Ramanujam, and D.C. Cooper. 1962. Cytogenetical studies in relation to the origin and differentiation of species in the genus *Solanum* L. Caryologia 15: 151–252.

Maguire, M.P. 1966. The relationship of crossing over to chromosome synapsis in a short paracentric inversion. Genetics 53: 1071–1077.

Maki, M., M. Masuda, and K. Inoue. 1996. Tetrasomic segregation of allozyme markers in an endangered plant, *Aster kantoensis.* J. Hered. 87: 378–380.

Manton, I. 1935. The cytological history of watercress (*Nasturtium officinale*). R. Br. Z. Indukt. Abstammungs-Vererbungsl. 69: 132–157.

Manton, I. 1937. The problem of *Biscutella laevigata* L. II. The evidence from meiosis. Ann. Bot. 1: 439–462.

Marks, G.E. 1966. The origin and significance of intraspecific polyploidy: experimental evidence from *Solanum chacoense.* Evolution 20: 552–557.

Marsden, J.E., S.J. Schwager, and B. May. 1987. Single-locus inheritance in the tetraploid treefrog *Hyla versicolor* with an analysis of expected progeny ratios in tetraploid organisms. Genetics 116: 299–311.

Martel, E., D. De Nay, S. Siljak-Yakovlev, S. Brown, and A. Sarr. 1997. Genome size variation and basic chromosome number in pearl millet and fourteen related *Pennisetum* species. J. Hered. 88: 139–143.

Martinez, A. 1976. Chromosome behaviour in some diploid species of Mexican tradescantias and their hybrids. Pages 224–225 in K. Jones and P.E. Brandham, eds., Current Chromosome Research. Elsevier/North Holland, Amsterdam.

Martínez, M., T. Naranjo, M.C. Cuadrado, and C. Romero. 1996. Synaptic behavior of tetraploid *Triticum timopheevii.* Theor. Appl. Genet. 93: 1139–1144.

Mastenbroek, I., J.M.J. DeWet, and C.-Y. Lu. 1982. Chromosome behaviour in early and advanced generation of tetraploid maize. Caryologia 35: 463–470.

Masterson, J. 1994. Stomatal size in fossil plants: evidence for polyploidy in majority of flowering plants. Science 264: 421–424.

Masuelli, R.W., E.L. Camadro, and A.O. Mendiburu. 1992. 2n gametes in *Solanum commersonii* and the cytological methods of triplandroid formation in triploid hybrids of *Solanum commersonii* x *Solanum gourlayi*. Genome 35: 864–869.

Matzke, M.A., and A.J.M. Matzke. 1998. Polyploidy and transposons. Trends Ecol. Evol. 13: 241.

Mauer, J., J.M. Mayo, and K. Denford. 1978. Comparative ecophysiology of the chromosome races in *Viola adunca* J.E. Smith. Oecologia 35: 91–104.

Maul, G.G. 1977. The nuclear and cytoplasmic pore complex: structure, dynamics, distribution and evolution. Int. Rev. Cytol. Suppl. 6: 75–186.

Maurizio, A. 1954. Untersuchungen über die Nektarsekretion einiger polyploider Kulturpflanzen. Arch. Julius Klaus-Stift. Verbungsforsch. Sozialanthropol. 29: 340–346.

McArthur, E.D., and S.C. Sanderson. 1999. Cytogeography and chromosome evolution of subgenus *Tridentatae* of *Artemisia* (Asteraceae). Am. J. Bot. 86: 1754–1775.

McArthur, E.D., J. Mudge, R. van Buren, W.R. Anderson, S.C. Sanderson, and D.G. Babbel. 1998. Randomly amplified polymorphic DNA analysis (RAPD) of *Artemisia* subgenus *Tridentatae* species and hybrids. Great Basin Nat. 58: 12–27.

McClintock, B. 1967. Genetic systems regulating gene expression during development. Pages 84–112 in M. Locke, ed., Control Mechanisms as Developmental Processes. Academic Press, New York.

McClintock, B. 1984. The significance of the response of the genome to challenge. Science 226: 792–801.

McCollum, G.D. 1958. Comparative studies of chromosome pairing in natural and induced tetraploid *Dactylis*. Chromosoma 9: 571–605.

McCullagh, D. 1934. Chromosome morphology in Plantaginaceae. I. Genetics 16: 1–44.

McDonald, J.F. 1995. Transposable elements: possible catalysts of organismic evolution. Trends Ecol. Evol. 10: 123–126.

McGuire P.E., and J. Dvořák. 1982. Genetic regulation of heterogenetic chromosome pairing in polyploid species of the genus *Triticum* sensu lato. Can. J. Genet. Cytol. 24: 57–82.

McHale, N.A. 1983. Environmental induction of high frequency $2n$ pollen formation in *Solanum*. Can. J. Genet. Cytol. 25: 609–615.

McMurphy, L.M., and A.L. Rayburn. 1991. Genome size variation in maize populations selected for cold tolerance. Plant Breed. 106: 190–195.

McMurphy, L.M., and A.L. Rayburn. 1992. Chromosomal and cell size of cold tolerant maize. Theor. Appl. Genet. 84: 798–802.

Mehta, R.K., and M.S. Swaminathan. 1957. Studies on induced polyploidy in forage crops. Indian J. Genet. Plant. Breed. 17: 27–57.

Menancio-Hautea, D., C.A. Fatokun, L. Kumar, D. Danash, and N.D. Young. 1993. Comparative chromosome analysis of mungbean (*Vigna radiata* L. Wilczek) and cowpea (*V. unguiculata* L. Walpers) using RFLP mapping data. Theor. Appl. Genet. 86: 797–810.

Menzel, M.Y. 1962. Pachytene chromosomes of the intergeneric hybrid *Lycopersicon esculentum* × *Solanum lycopersicoides*. Am. J. Bot. 49: 605–615.

Menzel, M.Y., and M.S. Brown. 1955. Isolating mechanisms in hybrids of *Gossypium gossypioides*. Am. J. Bot. 42: 49–57.

Meyer, V.G. 1965. Cytoplasmic effects on anther numbers in interspecific hybrids of cotton. I. Hered. 56: 292–294.

Meyer, V.G. 1970. A facultative gymnosperm from an interspecific cotton hybrid. Science 169: 886–888.

Meyer, V.G. 1971. Cytoplasmic effects on anther numbers in interspecific hybrids of cotton. II. *Gossypium herbaceum* and *G. herknesii* J. Hered. 62: 77–78.

Meyer, V.G. 1972. Cytoplasmic effects on anther numbers in interspecific hybrids of cotton. III. *Gossypium longicalyx*. J. Hered. 63: 33–34.

Michaelis, P. 1928. Über die Experimentalle Erzeugung Heteroploider Pflanzen bei *Epilobium* und *Oenothera*. Biol. Zentralb. 48: 370–374.

Miller, J.M. 1976. Variation in populations of *Claytonia perfoliata* (Portulacaeae). Syst. Bot. 1: 20–34.

Miller, J.S., and D.L. Venable. 2000. Polyploidy and the evolution of gender dimorphism in plants. Science 289: 2335–2338.

Milo, J., A. Levy, D. Palevitch, and G. Ladizinsky. 1987. Thebaine content and yield in induced tetraploid and triploid plants of *Papaver bracteatum* Lindl. Euphytica 36: 361–367.

Ming, R., S.-C. Liu, Y.-R. Lin, J. da Salva, W. Wilson, D. Braga, A. van Deynze, T.F. Wenslaff, K.K. Wu, P.H. Moore, W. Burnquist, M.E. Sorrels, J.E. Irvine, and A.H. Paterson. 1998. Detailed alignment of *Saccharum* and *Sorghum* chromosomes: comparative organization of closely related diploid and polyploid genomes. Genetics 150: 1663–1682.

Mittelsten Scheid, O., L. Jakovleva, K. Afsar, J. Maluszynska, and J. Paskowski. 1996. A change of ploidy can modify epigenetic silencing. Proc. Natl. Acad. Sci. USA 93: 7114–7119.

Moav, J., R. Moav, and D. Zohary. 1968. Spontaneous morphological alterations of chromosomes in *Nicotiana* hybrids. Genetics 59: 57–63.

Mok, D.W.S., and S.J. Peloquin. 1975. Three mechanisms of 2n pollen formation in diploid potatoes. Can. J. Genet. Cytol. 17: 217–225.

Moody, M.E., L.D. Mueller, and D.E. Soltis. 1993. Genetic variation and random drift in autotetraploid populations. Genetics 134: 649–657.

Moore, K. 1963. The influence of climate on a population of tetraploid spring wheat. Hereditas 50: 269–305.

Moore, G., K.J. Devos, Z. Wang, and M.D. Gale. 1995. Grasses, line up and form a circle. Curr. Biol. 5: 737–739.

Mooring, J. 1958. A cytogenetic study of *Clarkia unguiculata*. I. Translocations. Am. J. Bot. 45: 233–242.

Mooring, J.S. 2001. Barriers to interbreeding in the *Eriophyllum lanatum* (Asteraceae, Helenieae) species complex. Am. J. Bot. 88: 285–312.

Morgan, W.G., H. Thomas, M. Evans, and M. Borrill. 1986. Cytogenetic studies of interspecific hybrids between diploid species of *Festuca*. Can. J. Genet. Cytol. 28: 921–925.

Mosquin, T. 1964. Chromosomal repatterning in *Clarkia rhomboidea* as evidence for post-Pleistocene change in distribution. Evolution 18: 12–25.

Mosquin, T. 1967. Evidence for autopolyploidy in *Epilobium angustifolium* (Onagraceae). Evolution 21: 713–719.

Mosquin, T., and E. Small. 1971. An example of parallel evolution in *Epilobium* (Onagraceae). Evolution 25: 678–682.

Mummenhoff, K., A. Franzke, and M. Koch. 1997. Molecular phylogenetics of *Thaspi* s.l. (Brassicaceae) based on chloroplast DNA restriction-site variation and sequences of the internal transcribed spacers of nuclear ribosomal DNA. Can. J. Bot. 75: 469–482.

Müntzing, A. 1930a. Über Chromosomenvermehrung in *Galeopsis*-kreuzungen und ihre phylogenetsiche Bedeutung. Hereditas 14: 153–172.

Müntzing, A. 1930b. Outlines to a genetic monograph of the genus *Galeopsis* with special reference to the nature and inheritance of partial sterility. Hereditas 13: 185–341.

Müntzing, A. 1936. The evolutionary significance of autopolyploidy. Hereditas 21: 263–378.

Müntzing, A. 1940. Further studies on apomixis and sexuality in *Poa*. Hereditas 26: 115–190.

Müntzing, A. 1951. Cyto-genetic properties and practical value of tetraploid rye. Hereditas 37: 17–84.

Murata, M., Y. Ogurea, and F. Motoyoshi. 1994. Centromeric repetitive sequences in *Arabidopsis thaliana*. Jpn. J. Genet. 69: 361–370.

Murray, B.G., and C.A. Williams. 1976. Chromosome number and flavonoid synthesis in *Briza* L. (Gramineae). Biochem. Genet. 14: 897–904.

Mursal, I., and J.E. Endrizzi. 1976. A reexamination of the diploid-like meiotic behavior of polyploid cotton. Theor. Appl. Genet. 47: 171–178.

Nagl, W. 1974. Mitotic cycle time in perennial and annual plants with various amounts of DNA and heterochromatin. Dev. Biol. 39: 342–346.

Nagl, W., and F. Ehrendorfer. 1974. DNA content, heterochromatin, mitotic index and growth in perennial and annual *Anthemideae*. Plant Syst. Evol. 123: 35–54.

Nakai, Y. 1977. Variation of esterase isozymes and some soluble proteins in diploids and their induced autotetraploids in plants. Jpn. J. Genet. 52: 171–181.

Narayan, R.K.J. 1987. Nuclear DNA changes, genome differentiation, and evolution in *Nicotiana*. Plant Syst. Evol. 157: 161–180.

Narayan, R.K.J. 1988. Evolutionary significance of DNA variation in plants. Evol. Trends Plants. 2: 121–130.

Narayan, R.K.J., and A. Durrant. 1983. DNA distribution in chromosomes of *Lathyrus* species. Genetica 61: 47–53.

Narayan, R.K.J., and F.S. McIntyre. 1989. Chromosomal DNA variation, genomic constraints and recombination in *Lathyrus*. Genetica 79: 45–52.

Narayan, R.K.J., and H. Rees. 1976. Nuclear DNA variation in *Lathyrus*. Chromosoma 54: 141–154.

Natali, L., A. Cavallini, G. Cionini, O. Sassoli, P.G. Cionni, and M. Durante. 1993. Nuclear DNA changes within *Helianthus annuus* L.: changes within single progenies and their relationships with plant development. Theor. Appl. Genet. 85: 506–512.

Natali, L., T. Giordani, E. Polizzi, C. Pugliesi, M. Fambrini, and A. Cavallini. 1998. Genomic alterations in the interspecific hybrid *Helianthus annuus* × *Helianthus tuberosus*. Theor. Appl. Genet. 97: 1240–1247.

Naumova, T.N., M.D. Hayward, and M. Wagenvoort. 1999. Apomixis and sexuality in diploid and tetraploid accessions of *Brachiaria decumbens*. Sex. Plant Reprod. 12: 43–52.

Navashin, M. 1933. Altern der Samen als Ursache von Chromosomenmutationen. Planta 20: 233–243.

Neelam, A., and R.J. Narayah. 1994. Studies on backcross progeny of *N. rustica* × *N. tabacum* interspecific hybrids. II. Genomic DNA analyses. Cytologia 59: 385–391.

Ness, B.D., D.E. Soltis, and P.S. Soltis. 1989. Autopolyploidy in *Heuchera micrantha* (Saxifragaceae). Am. J. Bot. 76: 614–626.

Newman, L.J. 1966. Bridge and fragment aberrations in *Podophyllum peltatum*. Genetics 53: 55–63.

Nichols, C. 1941. Spontaneous chromosomal aberrations in *Allium*. Genetics 26: 89–100.

Nishiyama, I. 1942. Studies on the artificial polyploid plants. VI. On the different growth of the diploid and tetraploid radish in winter season. J. Hort. Soc. Jpn. 13: 245–252.

Noggle, G.R. 1946. The physiology of polyploidy in plants. I. Review of the literature. Lloydia 9: 153–173.

Noguti, Y., H. Oka, and T. Otuka. 1940. Studies on the polyploidy of *Nicotiana* induced by the treatment with colchicine. II. Growth rate and chemical analysis of diploid and its autotetraploid in *Nicotiana rustica* and *N. tabacum*. Jpn. J. Bot. 10: 343–364.

Norrmann, G.A., C.L. Quarin, and B.L. Burson. 1989. Cytogenetics and reproductive behavior of different chromosome races in six *Paspalum* species. J. Hered. 80: 24–28.

Novak, S.J., D.E. Soltis, and P.S. Soltis. 1991. Ownbey's tragopogons: 40 years later. Am. J. Bot. 78: 1586–1600.

Nunney, L. 1991. The influence of age structure and fecundity on effective population size. Proc. R. Soc. Lond. B. 246: 71–76.

Nygren, A. 1948. Further studies in spontaneous and synthetic *Calamagrostis purpurea*. Hereditas 34: 113–134.

Ohno, S. 1970. Evolution by Gene Duplication. Springer-Verlag, New York.

Ohri, D. 1996. Genome size and polyploidy variation in the hardwood genus *Terminalia* (Combretaceae). Plant Syst. Evol. 200: 225–232.

Ohri, D. 1998. Genome size variation and plant systematics. Ann Bot. 82 (Suppl. A): 75–83.

Ohri, D., R.M. Fritsch, and P. Hanelt. 1998. Evolution of genome size in *Allium* (Alliaceae). Plant Syst. Evol. 210: 57–86.

Ohri, D., A. Kumar, and M. Pal. 1986. Correlations between 2C DNA values and habit in *Cassia*. Plant Syst. Evol. 153: 223–227.

Ohri, D., and M. Pal. 1991. The origin of chickpea (*Cicer arietinum* L.): karyotype and nuclear DNA amount. Heredity 66: 367–372.

Oka, H, S.C. Hseheh, and T.S. Huang. 1954. Studies on tetraploid rice. V. The behavior of chromosomes in tetraploid rice varieties and their hybrids. Jpn. J. Genet. 29: 205–214.

Olmstead, R., and J.D. Palmer. 1991. Chloroplast DNA and systematics of the Solanaceae. Pages 161–168 in J.G. Hawkes and R.N. Lester, M. Nee, and N. Estrada, eds., Royal Botanical Gardens, Kew.

Orgel, L., and F.H.C. Crick. 1980. Selfish DNA: the ultimate parasite. Nature 284: 604–607.

Ornduff, R. 1970. Cytogeography of *Nymphoides* (Menyanthaceae). Taxon 19: 715–719.

Ortiz, R., N. Vorsa, L.P. Bruederle, and T. Laverty. 1992. Occurrence of unreduced pollen in diploid blueberry species, *Vaccineum* sect. *Cyanococcus*. Theor. Appl. Genet. 85: 55–60.

Osborne, D.J. 1982. Deoxyribonucleic acid integrity and repair in seed germination: the importance in viability and survival. Pages 435–463 in A.A. Khan, ed., The Physiology and Biochemistry of Seed Development, Dormancy, and Germination. Elsevier Biomedical, Amsterdam.

Otto, S.P., and J. Whitton. 2000. Polyploid incidence and evolution. Annu. Rev. Genet. 34: 401–437.

Ownbey, M. 1950. Natural hybridization and amphiploidy in the genus *Tragopogon*. Amer. J. Bot. 37: 487–499.

Ownbey, M. and G.D. McCollum. 1953. Cytoplasmic inheritance and reciprocal amphiploidy in *Tragopogon*. Amer. J. Bot. 40: 788–796.

Palmer, J.D., C.R. Shields, D.B. Cohen, and T.J. Horton. 1983. Chloroplast DNA evolution and the origin of amphidiploid *Brassica* species. Theor. Appl. Genet. 65: 181–189.

Palomino, G., G. Room, and S. Zarate. 1995. Chromosome numbers and DNA content in some taxa of *Leucaena* (Fabaceae, Mimosoideae). Cytologia 60: 31–37.

Pandey, K.K. 1955. Seed development in diploid, tetraploid and diploid-tetraploid crosses of *Trifolium pratense*. Indian J. Genet. Plant Breed. 15: 25–35.

Parker, J.S., and A.S. Wilby. 1989. Extreme chromosomal heterogeneity in a small-island population of *Rumex acetosa*. Heredity 62: 133–140.

Parks, C.R., and J.F. Wendel. 1990. Molecular divergence between Asian and North American species of *Liriodendron* (Magnoliaceae) with implications for the interpretation of fossil floras. Am. J. Bot. 77: 1243–1256.

Parokonny, A.S., A.E. Kenton, L. Meredith, S.J. Owens, and M.D. Bennett. 1992. Genomic divergence of allopatric sibling species studied by molecular cytogenetics of their F1 hybrids. Plant J. 2: 695–704.

Parrott, W.A., and R.R. Smith. 1984. Production of 2n pollen in red clover. Crop Sci. 24: 469–472.

Parrott, W.A., and R.R. Smith. 1986a. Current selection for 2n pollen formation in red clover. Crop Sci. 26: 1132–1135.

Parrott, W.A., and R.R. Smith. 1986b. Description and inheritance of new genes in red clover. J. Hered. 77: 355–358.

Parrott, W.A., and R.R. Smith. 1986c. Evidence for the existence of endosperm balance number in true clovers (*Trifolium* spp.). Can. J. Genet. Cytol. 28: 581–586.

Parsons, P.A. 1959. Some problems in inbreeding and random mating in tetrasomics. Agron. J. 51: 465–467.

Perez, F., A Mendendez, P. Dehal, and C.F. Quiros. 1999. Genomic structural differentiation in *Solanum:* comparative mapping of the A- and E-genomes. Theor. Appl. Genet. 98: 1183–1193.

Peters, A.D., and C.M. Lively. 2000. Epistasis and the maintenance of sex. Pages 99–112 in J.B. Wolf, E.D. Brodie III, and M.J. Wade, eds., Epistasis and the Evolutionary Process. Oxford University Press, New York.

Petit, C., F. Bretagnolle, and F. Felber. 1999. Evolutionary consequences of diploid-polyploid hybrid zones in wild species. Trends Ecol. Evol. 14: 306–311.

Petit, C., and J.D. Thompson. 1999. Species diversity and ecological range in relation to ploidy level in the flora of the Pyrenees. Evol. Ecol. 13: 45–66.

Peto, F.H. 1938. Cytology of poplar species and hybrids. Can. J. Res. 16: 445–455.

Pikaard, C.S. 2000. Nucleolar dominance: uniparental gene silencing on a multi-megabase scale in genetic hybrids. Plant. Mol. Biol. 43: 163–177.

Poggio, L., and J.H. Hunziker. 1986. Nuclear DNA content variation in *Bulnesia*. J. Hered. 77: 43–48.

Poggio, L., M.C. Molina, and C.A. Naranjo. 1990. Cytogenetic studies in the genus *Zea*. 2. Colchicine-induced multivalents. Theor. Appl. Genet. 79: 461–464.

Poggio, L., M. Rosato, A.M. Chiavarino, and C.A. Naranjo. 1998. Genome size and environmental correlations in maize (*Zea mays* ssp. *mays,* Poaceae). Ann. Bot. 82 (Suppl. A): 107–115.

Price, H.J., and K. Bachmann. 1975. DNA content and evolution in the Microseridae. Am. J. Bot. 62: 262–267.

Price, H.J., K.L. Chambers, and K. Bachmann. 1981a. Genome size variation in *Microseris biglovii* (Asteraceae). Bot. Gaz. 142: 156–159.

Price, H.J., K.L. Chambers, and K. Bachmann. 1981b. Geographic and ecological distribution of genomic DNA content variation in *Microseris douglasii* (Asteraceae). Bot. Gaz. 142: 415–426.

Price, H.J., K.L. Chambers, K. Bachmann, and J. Riggs. 1983. Inheritance of nuclear 2C DNA content variation in intraspecific and interspecific hybrids of *Microseris* (Asteraceae). Am. J. Bot. 70: 1133–1138.

Price, H.J., K.L. Chambers, K. Bachmann, and J. Riggs. 1986. Patterns of mean nuclear DNA content in *Microseris douglasii* (Asteraceae) populations. Bot. Gaz. 147: 496–507.

Price, H.J., K.L. Chambers, K. Bachmann, and J. Riggs. 1985. Inheritance of nuclear 2C DNA content in a cross between *Microseris douglasii* and *M. biglovii* (Asteraceae). Biol. Zbl. 104: 269–276.

Price, H.J., K.L. Chamber, and R.J. Bayer. 1984. Nuclear DNA contents of *Coreopsis nucensoides* and *C. nuecensis* (Asteraceae), a progenitor derivative pair. Bot. Gaz. 147: 496–507.

Pruitt, R.E., and E.M. Meyerowitz. 1986. Characterization of the genome of *Arabidopsis thaliana*. J. Mol. Biol. 187: 169–183.

Pustovoitova, T.N., and N.A. Borodina. 1981. Adaptive responses of polyploid plants under conditions of soil and atmospheric drought. Sov. Plant Physiol. (English Transl. Fiziol. Rast.) 28: 587–593.

Qu, L., J.F. Hancock, and J.H. Whallon. 1998. Evolution in an autotetraploid group displaying predominantly bivalent pairing at meiosis: genetic similarity of diploid *Vaccineum darrowi* and autotetraploid *V. corymbosum* (Ericaceae). Am. J. Bot. 85: 698–703.

Quarin, C.L., M.T. Pozzobon, and S.F.M. Wells. 1996. Cytological and reproductive behavior of diploid, tetraploid, and hexaploid germplasm accessions of a wild forage grass: *Paspalum compressifolium*. Euphytica 90: 345–349.

Quarin, C.L., G.A. Norrmann, and F. Espinoza. 1998. Evidence for autoploidy in apomictic *Paspalum rufum*. Hereditas 129: 119–124.

Quarin, C.L., F. Espinoza, E.J. Martinez, S.C. Pessino, and O.A. Bovo. 2001. A rise of ploidy level induces the expression of apomixis in *Paspalum notatum*. Sex. Plant Reprod. 13: 243–249.

Quillet, M.C., N. Madjidian, Y. Griveau, H. Serieya, M. Tersac, M. Lorieux, and A. Bervillé. 1995. Mapping genetic factors controlling pollen viability in an interspecific cross in *Helianthus* sect. *Helianthus*. Theor. Appl. Genet. 91: 1195–1202.

Quiros, C.F. 1982. Tetrasomic segregation for multiple alleles in alfalfa. Genetics 101: 117–127.

Rabinowitz, P.D. 2000. Are obese plant genomes on a diet? Genome Res. 10: 893–894.

Randhawa, A.S., and K.I. Beamish. 1970. Observation on the morphology, anatomy, classification, and reproductive cycle of *Saxifraga ferrugenia*. Can. J. Bot. 48: 299–312.

Rahn, K. 1957. Chromosome numbers in *Plantago*. Bot. Tidsskr. 53: 369–378.

Raina, S.N., A. Parida, K.K. Koul, S.S. Salimath, M.S. Bisht, V. Raja, and T.N. Khoshoo. 1994. Associated DNA changes in polyploids. Genome 37: 560–564.

Raina, S.N., and Y. Mukai. 1999. Genomic *in situ* hybridization in *Arachis* (Fabaceae) identifies the diploid wild progenitors of cultivated (*A. hypogaea*) and related wild (*A. monticola*) peanut species. Plant Syst. Evol. 214: 251–262.

Raina, S.N., and H. Rees. 1983. DNA variation between and within chromosome complements of *Vicia* species. Heredity 51: 335– 346.

Ramsey, J., and D.W. Schemske. 1998. Pathways, mechanisms, and rates of polyploid formation in flowering plants. Annu. Rev. Ecol. Syst. 29: 467–501.

Randolph, L.F. 1941. An evaluation of induced polyploids as a method of breeding crop plants. Am. Nat. 75: 347–363.

Randolph, L.F. 1942. Influence of heterozygosis on fertility and vigor in autotetraploid maize. Genetics 27: 163.

Raven, P.H. 1979. A survey of reproductive biology in the Onagraceae. NZ J. Bot. 17: 575–593.

Raven, P.H., W. Dietrich, and W. Stubbe. 1979. An outline of the systematics of *Oenothera* subsect. *Oenothera* (Onagraceae). Syst. Bot. 4: 242–252.

Raven, P.H., and D.P. Gregory. 1972. Observations of meiotic chromosomes in *Gaura* (Onagraceae). Brittonia 24: 71–86.

Raven, P.H., D.W. Kyhos, D.E. Breedlove, and W.W. Payne. 1968. Polyploidy in *Ambrosia dumosa* (Compositae: Ambrosiae). Brittonia 20: 205–211.

Raven, P.H., and D.R. Parnell. 1970. Two new species and nomenclatural changes in *Oenothera* subg. *Hartmannia* (Onagraceae). Madroño 20: 246–294.

Raven, P.H., O.T. Solbrig, D.W. Kyhos, and R. Snow. 1960. Chromosome numbers in the Compositae. I. Astereae. Am. J. Bot. 47: 124–132.

Rayburn, A.L., and J.A. Auger. 1990. Genome size variation of *Zea mays* ssp. *mays* adapted to different altitudes. Theor. Appl. Genet. 79: 470–474.

Rayburn, A.L., J.A. Auger, E.A. Benzinger, and A.G. Hepburn. 1985. C-band heterochromatin and DNA content in *Zea mays*. Am. J. Bot. 72: 1610–1617.

Rayburn, A.L., D.P. Biradar, D.G. Bullock, and L.M. McMurphy. 1993. Nuclear DNA content of F1 hybrids of maize. Heredity 70: 294–300.

Rayburn, A.L., J.W. Dudley, and D.P. Biradar. 1994. Selection for early flowering results in simultaneous selection for reduced nuclear DNA content in maize. Zeit. Pflanzenzucht. 112: 318–322.

Rees, H. 1961. Consequences of interchange. Evolution 15: 145–152.

Rees, H., and G.H. Jones. 1967. Chromosome evolution in *Lolium*. Heredity 22: 1–18.

Reeves, G., D. Frances, M.S. Davies, H.J. Rogers, and T R. Hodkinson. 1998. Genome size is negatively correlated with altitude in natural populations of *Dactylis glomerata*. Ann. Bot. 82 (Suppl. A): 99–105.

Reinisch, A.J., J.-M. Dong, C.L. Brubaker, D.M. Stelly, J.F. Wendel, and A.H. Paderson. 1994. A detailed map of cotton, *Gossypium hirsutum* × *Gossypium barbadense:* chromosome organization and evolution in a disomic polyploid genome. Genetics 138: 829–847.

Rhoades, M.M. 1951. Duplicate genes in maize. Am. Nat. 85: 105– 110.

Rhoades, M.W., and E. Dempsey. 1966. Induction of chromosome doubling at meiosis by the elongate gene in maize. Genetics 54: 505–522.

Rhyne, C.L. 1958. Linkage studies in *Gossypium*. I. Altered recombination in allotetraploid *G. hirsutum* L. following linkage group transference from a related diploid species. Genetics 43: 822–835.

Rice, W.R. 1994. Degeneration of a nonrecombining chromosome. Science 263: 230–232.

Rick, C.M. 1951. Hybrids between *Lycopersicon esculentum* Mill. and *Solanum lycopersicoides*. Proc. Natl. Acad. Sci. USA 37: 741–744.

Rick, C.M. 1969. Controlled introgression of chromosomes of *Solanum pennellii* into *Lycopersicon esculentum:* segregation and recombination. Genetics 62: 753–768.

Rieseberg, L.H. 1991. Homoploid reticulate evolution in *Helianthus* (Asteraceae): evidence from ribosomal genes. Am. J. Bot. 78: 1218–1237.

Rieseberg, L.H., M.A. Archer, and R.K. Wayne. 1999. Transgressive segregation, adaptation and speciation. Heredity 83: 363–372.

Rieseberg, L.H., D.M. Arias, M.C. Ungerer, C.R. Linder, and B. Sinervo. 1996. Role of gene interactions in hybrid speciation: evidence from ancient and experimental hybrids. Science 272: 741–745. Theor. Appl. Genet. 93: 633–644.

Rieseberg, L.H., S.M. Beckstrom-Sternberg, A. Liston, and D.M. Arias. 1991. Phylogenetic and systematic inferences from chloroplast DNA and isozyme variation in *Helianthus* (Asteraceae). Syst. Bot. 16: 50–76.

Rieseberg, L.H., C.R. Linder, and G.J. Seiler. 1995. Chromosomal and genic barriers to introgression in *Helianthus*. Genetics 141: 1163–1171.

Rieseberg, L.H., B. Sinervo, C.R. Linder, M.C. Ungerer, and D.M. Arias. 1996. Role of gene interactions in hybrid speciation: evidence from ancient and experimental hybrids. Science 272: 741–744.

Rieseberg, L.H., C. VanFossen, and A.M. Derochers. 1995. Hybrid speciation accompanied by genomic reorganization in wild sunflowers. Nature 375: 313–316.

Riley, R. 1960. The diploidization of polyploid wheat. Heredity 15: 407–429.

Riley, R. 1965. Cytogenetics and the evolution of wheat. Pages 102–122 in J. Hutchinson, ed., Essays on Crop Plant Evolution. Cambridge University Press, London.

Robertson, L.D., and K.J. Frey. 1984. Cytoplasmic effects on plant traits in interspecific matings of *Avena*. Crop. Sci. 24: 200–204.

Rodríguez, D.J. 1996. A model for the establishment of polyploidy on plants. Am. Nat. 147: 33–46.

Roelofs, D., J. van Velzen, P. Kuperus, and K. Bachmann. 1997. Molecular evidence for an extinct parent of the tetraploid species *Microseris acuminata* and *M. campestris* (Asteraceae, Lactuceae). Mol. Ecol. 6: 641–649.

Rohweder, H. 1937. Versuch zur erfassung der mengenmassigen Bedeckung des Dars und Zingst mit polyploiden Pflanzen. Planta 27: 501–549.

Rollins, R.C., and E.A. Shaw. 1973. The Genus *Lesquerella* (Cruciferae) in North America. Harvard University Press, Cambridge.

Roose, M.L., and L.D. Gottlieb. 1976. Genetic and biochemical consequences of polyploidy in *Tragopogon*. Evolution 30: 818–830.

Roose, M.L., and L.D. Gottlieb. 1978. Stability of structural gene number in diploid species with different amount of nuclear DNA and different chromosome numbers. Heredity 40: 159–163.

Rosser, E.M. 1955. A new British species of *Senecio*. Watsonia 3: 228.

Rothera, S.L., and A.J. Davy. 1986. Polyploidy and habitat differentiation in *Deschampsia caespitosa*. New Phytol. 10: 449–467.

Rothwell, N.V. 1959. Aneuploidy in *Claytonia virginica*. Am. J. Bot. 46: 353–360.

Rowe, D.A. 1986. Effects and control of genetic drift in the autotetraploid population. Crop Sci. 26: 89–92.

Rudolph, W., and P. Schwartz. 1951. Polyploidie effekte bei *Datura tatula*. Plant 39: 36–64.

Sakai, K. 1956. Studies on competition in plants. VI. Competition between autotetraploids and the diploid prototypes in *Nicotiana tabacum* L. Cytologia 21: 153–156.

Sakai, K., and Y. Suzuki. 1955a. Studies on competition in plants. II. Competition between diploid and autotetraploid plants of barley. J. Genet. 53: 11–20.

Sakai, K., and Y. Suzuki. 1955b. Studies on competition in plants. V. Competition between allopolyploids and their diploid parents. J. Genet. 53: 535–590.

Sakai, K., and H. Utiyamada. 1956. Studies on competition in plants. VIII. Chromosome number, hybridity, and competitive ability in *Oryza sativa* L. J. Genet. 54: 235–240.

Sanderson, S.C., and H.C. Stutz. 1994. High chromosome numbers in Mojavean and Sonoran Desert *Atriplex canescens* (Chenopodiaceae). Am. J. Bot. 81: 1045–1053.

Sandfaer, J. 1973. Barley stripe mosaic virus and the frequency of triploid and aneuploids in barley. Genetics 73: 597–603.

Sang, T., D.J. Crawford, and T.F. Stuessy. 1995. Documentation of reticulate evolution in peonies (*Paeonia*) using internal transcribed spacer sequences of nuclear ribosomal DNA: implications for biogeography and concerted evolution. Proc. Natl. Acad. Sci. USA 92: 6813–6817.

San Miguel, P., and J.L. Bennetzen. 1998. Evidence that recent increase in maize genome size was caused by massive amplification of intergene retrotransposons. Ann. Bot. 82 (Suppl. A): 37–44.

Sano, Y. 1980. Adaptive strategies compared between the diploid and tetraploid forms of *Oryza punctata*. Bot. Mag. Tokyo 93: 171–180.

Sapra, V.T., J.L. Hughes, and G.C. Sharma. 1975. Frequency, size and distribution of stomata in triticale leaves. Crop Sci. 15: 356–358.

Satina, S., and A.F. Blakeslee. 1935. Cytological effect of a gene in *Datura* which causes dyad formation in sporogenesis. Bot. Gaz. 96: 521–532.

Savchenko, V.K., and P.F. Rokitskii. 1975. Comparative study of the effectiveness of selection in diploid and autotetraploid populations. Genetika 11: 158–166.

Savidan, Y. 2000. Apomixis: genetics and breeding. Plant Breed. Rev. 18: 13–86.

Sax, K. 1931. Chromosome ring formation in *Rheo discolor*. Cytologia 3: 36–53.

Sax, K. 1933. Species hybrids in *Platanus* and *Campsis*. J. Arnold Arbor. 14: 274–278.

Sax, K. 1936. The experimental study of polyploidy. J. Arnold Arbor. 17: 153–159.

Schaefer, V.G., and J.P. Miksche. 1977. Microspectrophotometric determination of DNA per cell and polyploidy in *Fraxinus americana* L. Silvae Genet. 26: 184–192.

Schank, S.C, and P.F. Knowles. 1964. Cytogenetics of pairs of hybrids of *Carthamus* species (Compositae). Am. J. Bot. 51: 1093–1102.

Schilling, E.E., and C.B. Heiser. 1981. Intrageneric classification of *Helianthus* (Compositae). Taxon 30: 393–403.

Schilling, E.E., C.R. Linder, R.D. Noyes, and L.H. Rieseberg. 1998. Phylogenetic relationships in *Helianthus* (Asteraceae) based on nuclear ribosomal DNA internal transcribed spacer region sequence data. Syst. Bot. 23; 177–187.

Schluter, D., and L.M. Nagel. 1995. Parallel speciation by natural selection. Am. Nat. 146: 292–301.

Schmidt, T., and J.S. Heslop-Harrison. 1998. Genomes, genes and junk: the large scale organization of plant chromosomes. Trends Plant Sci. 3: 195–199.

Schmidt, T., and M. Metzlaff. 1991. Cloning and characterization of a *Beta vulgaris* satellite DNA family. Gene 101–247–250.

Schmidt, T., T. Schwartzacher, and J.S. Heslop-Harrison. 1994. Physical mapping of rRNA genes by fluorescent *in situ* hybridization and structural analysis of 5S rRNA genes and intergene spacer sequences in sugar beet (*Beta vulgaris*). Theor. Appl. Genet. 88: 629–636.

Schoen, D.J., and A.H.D. Brown. 1991. Intraspecific variation in population gene diversity and effective size correlates with the mating system in plants. Proc. Natl. Acad. Sci. USA 88: 4494–4497.

Schranz, M.E., and T.C. Osborn. 2000. Novel flowering time variation in the resynthesized polyploid *Brassica napus*. J. Hered. 91: 242–246.

Schröck, O. 1944. Untersuchungen an diploiden und tetraploiden Klonen von Luzerne, Gelbcress und Steinklee. Z. Pflanzenzücht. 26: 214–222.

Schubert, I., and R. Rieger. 1985. A new mechanism for altering chromosome number during karyotype evolution. Theor. Appl. Genet. 70: 213–221.

Schwanitz, F. 1957. Spornbildung bei einem Bastard zwischen drei *Digitalis*-Arten. Biol. Zentralbl. 76: 226–231.

Schwemmle, J. 1968. Selective fertilization in *Oenothera*. Adv. Genet. 14: 225–324.

Seal, A.G., and H. Rees. 1982. The distribution of quantitative DNA changes associated with the evolution of diploid *Festuceae*. Heredity 49: 179–190.

Sears, E.R. 1976. Genetic control of chromosome pairing in wheat. Annu. Rev. Genet. 10: 31–51.

Seavey, S.R., and P.H. Raven. 1977a. Chromosomal evolution in *Epilobium* sect. *Epilobium* (Onagraceae). Plant Syst. Evol. 127: 107–119.

Seavey, S.R., and P.H. Raven. 1977b. Chromosomal evolution in *Epilobium* sect. *Epilobium* (Onagraceae), II. Plant Syst. Evol. 128: 195–200.

Seberg, O., and G. Petersen. 1998. A critical review of concepts and methods used in classical genome analysis. Bot. Rev. 64: 373–417.

Seetharam, A. 1972. Interspecific hybridization in *Linum*. Euphytica 21: 489–495.

Segraves, K.A., and J.N. Thompson. 1999. Plant polyploidy and pollination: floral traits and insect visits to diploid and tetraploid *Heuchera grossularifolia*. Evolution 53: 1114–1127.

Setter, T.L., L.E. Schrader, and E.T. Bingham. 1978. Carbon dioxide exchange rates, transpiration and leaf characteristics in genetically equivalent ploidy levels in alfalfa. Crop Sci. 18: 327–332.

Shang, X.M., R.C. Jackson, H.T. Nguyen, and J.Y. Huang. 1989. Chromosome pairing in the *Triticum monococcum* complex: evidence for pairing control genes. Genome 32: 216–226.

Sharma, A.K., and D. Dey. 1967. A comprehensive cytotaxonomic study on the family Chenopodiaceae. J. Cytol. Genet. 2: 114–127.

Shi, L., T. Zhu, and P. Keim. 1996. Ribosomal RNA genes in soybean and common bean: chromosome organization, expression and evolution. Theor. Appl. Genet. 93: 136–141.

Shoemaker, R., T. Olsen, and V. Kanazin. 1996. Soybean genome organization: evolution of a legume genome. Pages 139–150 in J.P. Gustafson and R.B. Flavell, eds., Genomes of Plant and Animals: 21st Stadler Genetics Symposium. Plenum Press, New York.

Shore, J.S. 1991a. Chromosomal evidence of autopolyploidy in the *Turnera ulmifolia* complex (Turneraceae). Can. J. Bot. 69: 1302–1308.

Shore, J.S. 1991b. Tetrasomic inheritance and isozyme variation in *Turnera ulmifolia* var. *elegans* UIB. and *intermedia* UIB. (Turneraceae). Heredity 66: 305–312.

Sieber, V.K. and B.G. Murray. 1981. Structural and numerical chromosomal polymorphism in natural populations of *Alopecurus* (Poaceae). Plant Syst. Evol. 139: 121–136.

Simonsen, O. 1973. Cytogenetic investigations in diploid and autotetraploid populations of *Lolium perenne* L. Hereditas 75: 157–188.

Simonsen, O. 1975. Cytogenetic investigations in diploid and tetraploid populations of *Festuca pratensis*. Hereditas 79: 73–108.

Sims, L.E., and H.J. Price. 1985. Nuclear DNA content variation in *Helianthus* (Asteraceae). Am. J. Bot. 72: 1213–1219.

Singh, A.K. 1986. Utilization of wild relatives in the genetic improvement of *Arachis hypogaea* L. 7. Autotetraploid production and prospects of interspecific interbreeding. Theor. Appl. Genet. 72: 164–169.

Singh, K.P., S.N. Raina, and A.K. Singh. 1996. Variation in chromosomal DNA associated with the evolution of *Arachis* species. Genome 39: 890–897.

Singh, R.J., K.P. Kollipara, and T. Hymowitz. 1987. Polyploid complexes of *Glycine tabacina* (Labill.) Benth. and *G. tomentella* Hayata revealed by cytogenetic analysis. Genome 29: 490–497.

Small, E. 1968. The systematics of autopolyploidy in *Epilobium latifolium* (Onagraceae). Brittonia 20: 169–181.

Smith, D.M., and D.A. Levin. 1963. A chromatographic study of reticulate evolution in the Appalachian *Asplenium* complex. Am. J. Bot. 50: 952–958.

Smith, E.B. 1966. Cytogenetics and phylogeny of *Haplopappus* section *Isopappus* (Compositae). Can. J. Genet. Cytol. 8: 14–36.

Smith, E.B. 1974. *Coreopsis nuecensis* (Compositae) and a related new species from southern Texas. Brittonia 26: 161–171.

Smith, E.C. 1941. Chromosome behavior in *Catalpa hybrida* Spathe. J. Arnold Arbor. 22: 219–221.

Smith, H.E. 1946. *Sedum pulchellum:* a physiological and morphological comparision of diploid, tetraploid and hexaploid races. Bull. Torrey Bot. Club. 73: 495–541.

Smith, J.B., and M.D. Bennett. 1975. DNA variation in the genus *Ranunculus*. Heredity 35: 231–239.

Smith-Huerta, N.L. 1986. Isozyme diversity in three allotetraploid *Clarkia* species and their putative progenitors. J. Hered. 77: 349–354.

Snogerup, S. 1967. Studies in the Aegean flora. IX. *Erysimum* sect. *Cheiranthus*. B. Variation and evolution in the small population system. Opera Bot. 14: 1–86.

Snow, R. 1960. Chromosomal differentiation in *Clarkia dudleyana*. Am. J. Bot. 47: 302–309.

Snow, R. 1963. Cytogenetic studies in *Clarkia*, section *Primigenia*. I. A cytological survey of *Clarkia amoena*. Am. J. Bot. 50: 337–348.

Snowden, R.J., W. Köhler, and A. Köhler. 1997. Chromosomal localization and characterization of rDNA loci in the *Brassica* A and C genomes. Genome 40: 852–857.

Solbrig, O. 1968. Fertility, sterility and the species problem. Pages 77–96 in V.H. Heywood, ed., Modern Methods in Plant Taxonomy. Academic Press, London.

Soltis, D.E. 1984. Autopolyploidy in *Tolmiea menziesii*. (Saxifragaceae). Am. J. Bot. 71: 1171–1174.

Soltis, D.E., and J.J. Doyle. 1987. Ribosomal RNA gene variation in diploid and tetraploid *Tolmiea menziesii* (Saxifragaceae). Biochem. Syst. Ecol. 15: 75–78.

Soltis, D.E., and P.S. Soltis. 1989a. Genetic consequences of autopolyploidy in *Tolmiea* (Saxifragaceae). Evolution 43: 586–594.

Soltis, D.E., and P.S. Soltis. 1989b. Allopolyploid speciation in *Tragopogon:* insights from chloroplast DNA. Am. J. Bot. 76: 1119–1124.

Soltis, D.E., and P.S. Soltis. 1989c. Tetrasomic inheritance in *Heuchera micrantha* Saxifragaceae. J. Hered. 80: 123–126.

Soltis, D.E., and P.S. Soltis. 1993. Molecular data and the dynamic nature of polyploidy. Crit. Rev. Plant. Sci. 12: 243–273.

Soltis, D.E., and P.S. Soltis. 1999. Polyploidy: recurrent formation and genome evolution. Trends Ecol. & Evol. 14: 348–352.

Soltis, P.S., and W.L. Bloom. 1991. Allozymic differentiation between *Clarkia nitens* and *C. speciosa* (Onagraceae). Syst. Bot. 16: 399–406.

Soltis, P.S., G.M. Plunkett, S.J. Novak, and D.E. Soltis. 1995. Genetic variation in *Tragopogon* species: additional origins of the allotetraploids *T. mirus* and *T. miscellus* (Compositae). Am. J. Bot. 82: 1329–1341.

Somaroo, B.H., and W.F. Grant. 1971. Meiotic chromosome behavior in induced autotetraploids and amphidiploids in the *Lotus corniculatus* group. Can. J. Genet. Cytol. 13: 663–671.

Song, K., P. Lu, K. Tang, and T. Osborn. 1995. Rapid genome change in synthetic polyploids of *Brassica* and its implications for polyploid evolution. Proc. Natl. Acad. Sci. USA 92: 7719–7723.

Song, K., and T.C. Osborn. 1992. Polyphyletic origins of *Brassica napus:* new evidence based on organelle and nuclear RFLP analyses. Genome 35: 992–1001.

Song, K., T.C. Osborn, and P.H. Williams. 1988. *Brassica* taxonomy based on nuclear restriction length fragment polymorphisms (RFLPs). 1. Genome evolution of diploid and amphidiploid species. Theor. Appl. Genet. 75: 784–794.

Spellenberg, R. 1981. Polyploidy in *Dalea formosa* (Fabaceae) on the Chihuahuan Desert. Brittonia 33: 309–324.

Srivastava, S., and U.C. Lavania. 1991. Evolutionary DNA variation in *Papaver.* Genome 34: 763–768.

Srivastava, S., U.C. Lavania, and J. Sybenga. 1992. Genetic variation and meiotic behaviour and fertility in tetraploid *Hyocyamus muticus:* correlation with diploid meiosis. Heredity 68: 231–239.

Stace, H.M. 1978. Cytoevolution in the genus *Calotis* R. Br. (Compositae: Astereae). Austral. J. Bot. 26: 287–307.

Stace, H.M. 1982. *Calotis* (Compositae), a Pliocene arid zone genus? Pages 357–367 in W.R. Barker and P.J.M. Greenslade, eds., Evolution of the Flora of Arid Australia. Peacock Publications, Frewirlle, Australia.

Stace, H.M., and S.H. James. 1996. Another perspective on cytoevolution in Lobelioideae (Campanulaceae). Am. J. Bot. 83: 1356–1364.

Stahevitch, A.E., and W.A. Wojtas. 1988. Chromosome numbers of some North American species of *Artemisia* (Asteraceae). Can. J. Bot. 66: 672–676.

Stebbins, G.J. 1939. Notes on the systematic relationships of the Old World species and of some horticultural forms of *Paeonia.* Univ. Calif. Publ. Bot. 19: 245–266.

Stebbins, G.J. 1947. Types of polyploids: their classification and significance. Adv. Genet. 1: 403–429.

Stebbins, G.J. 1949. The evolutionary significance of natural and artificial polyploids in the family Gramineae. Hereditas 36: (Suppl.): 461–485.

Stebbins, G.J. 1950. Variation and Evolution in Plants. Columbia University Press, New York.

Stebbins, G.J. 1956. Artificial polyploidy as a tool in plant breeding. Brookhaven Symp. Biol. 9: 37–50.

Stebbins, G.J. 1958a. Longevity, habitat, and release of genetic variability in higher plants. Cold Spring Harbor Symp. Quant. Biol. 23: 365–378.

Stebbins, G.J. 1966. Chromosome variation and evolution. Science 152: 1563–1469.

Stebbins, G.L. 1971. Chromosomal Evolution in Higher Plants. Addison-Wesley, London.

Stebbins, G.J. 1972. Research on the evolution of higher plants: problems and prospects. Can. J. Genet. Cytol. 14: 453–462.

Stebbins, G.J. 1976. Chromosomes, DNA, and plant evolution. Evol. Biol. 9: 1–34.

Stebbins, G.J. 1984. Polyploidy and the distribution of the arctic-alpine flora: new evidence and a new approach. Bot. Helv. 94: 1–13.

Stebbins, G.J. 1985. Polyploidy, hybridization, and the invasion of new habitats. Ann. Missouri Bot. Gard. 72: 824–832.

Stebbins, G.J., and J.C. Dawe. 1987. Polyploidy and distribution in the European flora: a reappraisal. Bot. Jahrb. Syst. 108: 343–354.

Stebbins, G.J., and S. Ellerton. 1939. Structural hybridity in *Paeonia californica* and *P. brownii*. J. Genet. 38: 1–36.

Stebbins, G.J., and F.T. Pun. 1953. Artificial and natural hybrids in the Gramineae. V. Diploids of *Agropyron*. Am. J. Bot. 40: 444–449.

Stedje, B. 1989. Chromosome evolution within the *Ornithogalum tenuifolium* complex (Hyacinthaceae), with special emphasis on the evolution of bimodal karyotypes. Plant Syst. Evol. 166: 79–89.

Steiner, E. 1956. New aspects of the balanced lethal system in *Oenothera*. Genetics 41: 487–500.

Steiner, E. 1960. Incompatibility in the complex heterozygotes of *Oenothera*. Genetics 46: 301–305.

Steiner, E. 1974. *Oenothera*. Pages 233–245 in R.C. King, ed., Handbook of Genetics, vol. 2. Plenum, New York.

Steiner, E., and D.A. Levin. 1977. Allozyme, SI gene, cytological and morphological polymorphisms in a population of *Oenothera biennis*. Evolution 31: 127–133.

Stephens, S.G. 1961. Species differentiation in relation to crop improvement. Crop Sci. 1: 1–4.

Stevens, J.P., and S.M. Bougourd. 1991. The frequency and meiotic behaviour of structural chromosome variants in natural populations of *Allium schoenoprasum* L. (wild chives) in Europe. Heredity 66: 391–401.

Stoilov, M., G. Jannson, G. Eriksson, and L. Ehrenberg. 1966. Genetical and physiological causes of the variation of radiosensitivity in barley and maize. Radiat. Bot. 6: 457–467.

Stout, A.B., and C. Chandler. 1942. Hereditary transmission of induced polyploidy and compatibility in fertilization. Science 96: 257.

Stoutamire, W.P. 1977. Chromosome races of *Gaillardia pulchella* (Asteraceae). Brittonia 29: 297–309.

Strid, A. 1970. Studies in the Aegean flora. XVI. Biosystematics of the *Nigella arvensis* complex with special reference to the problem of non-adaptive radiation. Opera Bot. no. 28.

Stubbe, W. 1980. Über die Bedingungen der Komoplex heterozygotie und die Beiden Wegen der Evolutionen komplexheterozygotischer Arten bei *Oenothera*. Ber. Dtsch. Bot. Ges. 93: 441–447.

Stubbe, W., and P.H. Raven. 1979. A genetic contribution to the taxonomy of *Oenothera* Sect. *Oenothera* (including subsections *Oenothera, Emersonia, Raimannia,* and *Munzia*). Plant Syst. Evol. 133: 39–59.

Stuessy, T.F., and D.J. Crawford. 1998. Chromosomal stasis during speciation in angiosperms of oceanic islands. Pages 307–324 in T. Stuessy and M. Ono, eds., Evolution and Speciation of Island Plants. Cambridge University Press, Cambridge.

Stutz, H.C., and S.C. Sanderson. 1983. Evolutionary studies of *Atriplex:* chromosome races of *A. confertifolia* (shadscale). Am. J. Bot. 70: 1536–1547.

Swaminathan, M.S., and H.W. Howard. 1953. The cytology and genetics of the potato (*Solanum tuberosum*) and related species. Bibliogr. Genet. 14: 1–192.

Swietlinska, Z., B. Lotocka-Jakubowska, and J. Zuk. 1971. Cytogenetical relationships between *Rumex tuberosus, R. thrysiflorus* and *R. acetosa* and occurrence of polyploidy among their hybrids. Theor. Appl. Genet. 41: 150–156.

Sybenga, J. 1975. Meiotic Configurations. Springer-Verlag, New York.

Sybenga, J. 1992. Cytogenetics in Plant Breeding. Springer-Verlag, New York.

Sybenga, J. 1996. Chromosome pairing affinity and quadrivalent formation in polyploids: do segmental polyploids exist? Genome 39: 1176–1184.

Tadmor, Y., D. Zamir, and G. Ladizinsky. 1987. Genetic mapping of an ancient translocation in the genus *Lens*. Theor. Appl. Genet. 73: 883–892.

Taketa, S., H. Ando, K. Taketa, and R. von Bothmer. 1999. Detection of *Hordeum marinum* genome in three polyploid *Hordeum* species and cytotypes by genomic in situ hybridization. Hereditas 130: 185–188.

Taketa, S., G.E. Harrison, and J.S. Heslop-Harrison. 1999. Comparative physical mapping of the 5S and 18S-25S rDNA in nine wild *Hordeum* species and cytotypes. Theor. Appl. Genet. 98: 5–9.

Tal, M. 1980. Physiology of polyploids. Pages 61–75 in W.H. Lewis, ed., Polyploidy—Biological Relevance. Plenum Press, New York.

Tanksley, S.D., M.W. Ganal., J.P. Price, M.C. De Vicente, M.W. Bonierbale, P. Broun, T.M. Fulton, J.J. Giovannoni, S. Grandillo, G.B. Martin, R. Messeguer, J.C. Miller, L. Miller, A.H. Paterson, O. Pineda, M.S. Röder, R.A. Wing, W. Wu, and N.D. Young. 1992. High density molecular linkage maps of the tomato and potato genomes. Genetics 132: 1141–1160.

Tavoletti, S., A. Mariani, and F. Veronesi. 1991. Phenotypic recurrent selection for 2n pollen and 2n egg production in diploid alfalfa. Euphytica 57: 97–102.

Taylor, I.B., and G.M. Evans. 1977. The genotypic control of homoeologous chromosome association in *Lolium temulentum* × *Lolium perenne* interspecific hybrids. Chromosoma 62: 57–67.

Taylor, N.L., and R.R. Smith. 1979. Red clover breeding and genetics. Adv. Agron. 31: 125–154.

Templeton, A.R. 1981. Mechanisms of speciation—a population genetic approach. Annu. Rev. Ecol. Syst. 12: 23–48.

Templeton, A.R., and D.A. Levin. 1979. Evolutionary consequences of seed pools. Am. Nat. 114: 232–249.

Thomas, H.M., and W.G. Morgan. 1990. Analyses of synaptonemal complexes and chromosome pairing at metaphase-I in the diploid intergeneric hybrid × *Festuca drymeja*. Genome 33: 465–471.

Thomas, H.M., and R.A. Pickering. 1985. The influence of parental genotype on the chromosome behaviour of *Hordeum vulgare* × *H. bulbosum* diploid hybrids. Theor. Appl. Genet. 71: 437–442.

Thomas, H.M., and B.J. Thomas. 1993. Synaptonemal complex formation in two allohexaploid *Festuca* species and a pentaploid hybrid. Heredity 71: 305–311.

Thomas, J.H. 1993. Thinking about redundancy. Trends Genet. 9: 395–399.

Thompson, J.D., and R. Lumaret. 1992. The evolutionary dynamics of polyploid plants: origins, establishment and persistence. Trends Ecol. Evol. 7: 302–307.

Thompson, J.N., B.M. Cunningham, K.A. Seagraves, D.M. Althoff, and D. Wagner. 1997. Plant polyploidy and insect/plant interactions. Am. Nat. 150: 730–743.

Thompson, K. 1990. Genome size, seed size, and germination temperature in herbaceous angiosperms. Evol. Trends Plants 4: 113–116.

Togby, H.A. 1943. A cytological study of *Crepis fuliginosa, C. neglecta,* and their F1 hybrid, and its bearing on the mechanism of phyletic reduction in chromosome number. J. Genet. 45: 67–111.

Tomekpe, K., and R. Lumaret. 1991. Association between quantitative traits and allozyme heterozygosity in a tetrasomic species: *Dactylis glomerata.* Evolution 45: 359–370.

Towner, H.R. 1977. The biosystematics of *Calylophus* (Onagraceae). Ann. Missouri Bot. Gard. 64: 48–120.

Townsend, C.E., and E.E. Remmenga. 1968. Inbreeding in tetraploid alsike clover, *Trifolium hybridum* L. Crop Sci. 8: 213–217.

Tsay, Y.-F, M.J. Frank, T. Page, C. Dean, and N.M. Crawford. 1993. Identification of a mobile endogenous transposon in *Arabidopsis thaliana.* Science 260: 342–344.

Turpeinen, T., J. Kulmala, and E. Nevo. 1999. Genome size variation in *Hordeum spontaneum* populations. Genome 42: 1094–1099.

Tyagi, B.R. 1988. The mechanism of 2n pollen formation in diploids of *Costus speciosus* (Koenig) J.E. Smith and the role of sexual polyploidization in the origin of intraspecific chromosomal races. Cytologia 53: 763–770.

Tyler, B., M. Borrill, and K. Chorlton. 1978. Studies in *Festuca* X. Observations on germination and seedling cold tolerance in diploid *Festuca pratensis* and tetraploid *F. pratensis* var. *apennina* in relation to their altitudinal distribution. J. Appl. Ecol. 15: 219–226.

U, N. 1935. Genome-analysis in *Brassica* with special reference to the experimental formation of *B. napus* and peculiar mode of fertilization. Jpn. J. Bot. 7: 389–452.

Valentine, D.H. 1952. Studies in British *Primula* III. Hybridization between *Primula elatior* (L.) Hill and *P. veris* L. New Phytol. 50: 383–398.

Van Dijk, P., and T. Bakx-Schotman. 1997. Chloroplast DNA phylogeography and cytotype geography in autotetraploid *Plantago media.* Mol. Ecol. 6: 345–352.

Van Dijk, P., M. Hartog, and W. Van Delden. 1992. Single cytotype areas in autotetraploid *Plantago media* L. Biol. J. Linn. Soc. 46: 315–331.

van Houten, W.H.J., N. Scarlett, and K. Bachmann. 1993. Nuclear DNA markers of the Australian tetraploid *Microseris scapigera* and its North American diploid relatives. Theor. Appl. Genet. 87: 498–505.

Vasek, F.C. 1960. A cytogenetic study of *Clarkia exilis.* Evolution 14: 88–97.

Vasiléva, M.G., E.V. Kljujkov, and M.G. Pimenov. 1985 Karyotaxonomic analyses of the genus *Bunium* Umbelliferae. Plant Syst. Evol. 149: 71–88.

Veilleux, R.E., and F.I. Lauer. 1981. Variation for 2n pollen production in clones of *Solanum phureja* Juz. and Buk. Theor. Appl. Genet. 59: 95–100.

Verma, S.C., and H. Rees. 1974. Nuclear DNA and the evolution of allotetraploid Brassiceae. Heredity 33: 61–68.

Vestad, R. 1960. The effect of induced polyploidy on resistance to clover rot (*Sclerotinia trifoliorum* Erikss.) in red clover. Euphytica 9: 35–38.

Vinogradov, A.E. 1999. Intron-genome size relationship on a large evolutionary scale. J. Mol. Evol. 49: 376–384.

Volkov, R.A., N.V. Borisjuk, I.I. Panchuk, D. Schweitzer, and V. Hemleben. 1999. Elimination and rearrangement of parental rDNA in the allotetraploid *Nicotiana tabacum*. Mol. Biol. Evol. 16: 311–320.

Vrijenhoek, R.C. 1984. Ecological differentiation among clones: the frozen niche variation model. Pages 217–231 in K. Wöhrmann and V. Loeschcke, eds., Population Biology and Evolution. Springer-Verlag, Berlin.

Wagenaar, E.B. 1970. Studies on the genome constitution of *Triticium timopheevi* Zhuk. III. Segregation of meiotic chromosome behavior in backcross generations. Can. J. Genet. Cytol. 12: 347–355.

Wagner, W.H., Jr. 1954. Reticulate evolution in the Appalachian aspleniums. Evolution 8: 103–118.

Wain, R.P. 1983. Genetic differentiation during speciation in the *Helianthus debilis* complex. Evolution 37: 1119–1127.

Walbot, V., and C.A. Cullis. 1985. Rapid genome changes in higher plants. Annu. Rev. Plant Physiol. 36: 367–396.

Walbot, V., and D.A. Petrov. 2001. Gene galaxies in the maize genome. Proc. Natl. Acad. Sci. USA 98: 8163–8164.

Wall, J.R. 1970. Experimental introgression in the genus *Phaseolus*. I. Effect of mating systems on interspecific gene flow. Evolution 24: 356–366.

Walsh, J.B. 1982. Rate of accumulation of reproductive isolation by chromosome rearrangements. Amer. Natur. 120: 510–532.

Walsh, J.B. 1985. How many processed pseudogenes are accumulated in a gene family? Genetics 110: 354–364.

Walsh, J.B. 1995. How often do duplicated genes evolve new functions? Genetics 139: 421–428.

Walters, J.L. 1942. Distribution of structural hybrids in *Paeonia californica*. Am. J. Bot. 29: 270–275.

Walters, J.L. 1952. Heteromorphic chromosome pairs in *Paeonia californica*. Am. J. Bot. 39: 145–151.

Walters, J.L. 1956. Spontaneous meiotic chromosome breakage in natural populations of *Paeonia californica*. Am. J. Bot. 43: 342–353.

Wang, G.-L., J.-M. Dong, and A.H. Paterson. 1995. The distribution of *Gossypium hirsutum* chromatin in *G. barbadense* germplasm: molecular analysis of introgressive breeding. Theor. Appl. Genet. 91: 1153–1161.

Wang, G.-Z., N.T. Miyashita, and K. Tsunewaki. 1997. Plasmon analyses of *Triticum* (wheat) and *Aegilops:* PCR-single-strand conformational polymorphism (PCR-SSCP) analyses of organellar DNAs. Proc. Natl. Acad. Sci. USA 94: 14570–14577.

Wang, T., W. Liu, C. He, G. Zhang, R. Shi, X. Xu, and J. Li. 1999. Effect of alien cytoplasm on meiosis in the F1 hybrids between alloplasmic lines of *Triticum aestivum* cv. Chinese Spring and rye or triticale. Cytologia 64: 285–291.

Warmke, H.E., and A.F. Blakeslee. 1940. The establishment of a 4*n* dioecious race in *Melandrium*. Am. J. Bot. 27: 751–762.

Warner, D.A., and G.E. Edwards. 1988. C4 photosynthesis and leaf anatomy in diploid and tetraploid *Pennisetum americanum* (pearl millet). Plant Sci. 56: 85–92.

Warner, D.A., and G.E. Edwards. 1989. Effects of polyploidy on photosynthetic rates, photosynthetic enzymes, contents of DNA, chlorophyll, and sizes and numbers of photosynthetic cells in the C4 dicot *Atriplex confertifolia.* Plant Physiol. 91: 1143–1151.

Warner, D.A., M.S.B. Ku, and G.E. Edwards. 1987. Photosynthesis, leaf anatomy, and cellular constituents in the polyploid C4 grass *Panicum virgatum.* Plant Physiol. 84: 461–466.

Wasmund, O., and W. Stubbe. 1986. Cytogenetic investigations on *Oenothera wolfii* (Onagraceae). Plant Syst. Evol. 154: 79–88.

Watanabe, K. 1983. Studies on the control of diploid-like meiosis in polyploid taxa of *Chrysanthemum.* Theor. Appl. Genet. 66: 9–14.

Watanabe, K. 1986. The cytogeography of the genus *Eupatorium* (Compositae). A review. Plant Species Biol. 1: 99–116.

Watanabe, K., and S.J. Peloquin. 1991. Occurrence and frequency of 2n pollen in 2x, 4x, and 6x wild tuber-bearing species from Mexico, and Central and South America. Theor. Appl. Genet. 82: 621–626.

Watanabe, K., T. Yahara, T. Denda, and K. Kosuge. 1999. Chromosomal evolution in the genus *Brachyscome* (Asteraceae, Astereae): statistical tests regarding correlation between changes in karyotype and habitat using phylogenetic information. J. Plant Res. 112: 145–161.

Waters, E.R., and B.A. Schaal. 1996. Heat shock indices a loss of r-RNA-encoding DNA repeats in *Brassica nigra.* Proc. Natl. Acad. Sci. USA 93: 1449–1452.

Wedberg, H.L., H. Lewis, and C.S. Venkatesh. 1968. Translocation heterozygotes and supernumerary chromosomes in wild populations of *Clarkia williamsonii.* Evolution 22: 93–107.

Weiss, R.L., J.R. Kukora, and J. Adams. 1975. The relationship between enzyme activity, cell geometry, and fitness in *Saccharomyces cerevisiae.* Proc. Natl. Acad. Sci. USA 72: 794–798.

Welch, D.M., and M. Meselson. 2000. Evidence for the evolution of bdelloid rotifers without sexual reproduction or genetic exchange. Science 288: 1211–1214.

Welch, D.M., and M. Meselson. 2001. Rates of nucleotide substitution in sexual and anciently asexual rotifers. Proc. Natl. Acad. Sci. USA 98: 6720–6724.

Wendel, J.F. 1989. New World cottons contain Old World cytoplasm. Proc. Natl. Acad. Sci. USA 86: 4132–4136.

Wendel, J.F. 2000. Genome evolution in polyploids. Plant Mol. Biol. 42: 225–249.

Wendel, J.F., and V.A. Albert. 1992. Phylogenetics of the cotton genus (*Gossypium*): character-weighted parsimony analysis of the chloroplast-DNA restriction site data and its systematic and biogeographic implications. Syst. Bot. 17: 115–143.

Wendel, J.F., A. Schnabel, and T. Seelanan. 1995a. Bidirectional concerted evolution following allopolyploid speciation in cotton (*Gossypium*). Proc. Natl. Acad. Sci. USA 92: 280–284.

Wendel, J.F., A. Schnabel, and T. Seelanan. 1995b. An unusual ribosomal DNA sequence from *Gossypium gossypioides* reveals ancient, cryptic, intergenomic introgression. Mol. Phylogenet. Evol. 4: 298–313.

Wendel, J.F., C.W. Stuber, M.D. Edwards, and M.M. Goodman. 1986. Duplicated chromosome segments in *Zea mays* L.: further evidence from hexokinase isozymes. Theor. Appl. Genet. 72: 178–185.

Wendel, J.F., C.W. Stuber., M.M. Goodman, and J.B. Beckett. 1989. Duplicated plastid and triplicated cytosolic isozymes of triosephosphate isomerase in maize (*Zea mays* L.). J. Hered. 80: 218–228.

Werner, J.E., and S.J. Peloquin. 1991. Occurrence and mechanisms of $2n$ egg formation in $2x$ potato. Genome 34: 975–982.

Werth, C.R., S.I. Guttman, and W.H. Eshbaugh. 1985. Electrophoretic evidence of reticulate evolution in the Appalachian *Asplenium* complex. Syst. Bot. 10: 184–192.

Wettsein, D. von, S.W. Rasmussen, and P.B. Holm. 1984. The synaptonemal complex in genetic segregation. Annu. Rev. Genet. 18: 331–413.

Whitkus, R., J. Doebley, and M. Lee. 1992. Comparative genome mapping of sorghum and maize. Genetics 132: 1119–1130.

Whittingham, A.D., and G.J. Stebbins. 1969. Chromosomal rearrangements in *Plantago insularis*. Eastw. Chromosoma 26: 449–468.

Williams, C.G., M.M. Goodman, and C.W. Stuber. 1995. Comparative recombination distances among *Zea mays* L. inbreds, wide crosses and interspecific hybrids. Genetics 141: 1573–1581.

Wilson, H.S., S.C. Barber, and T. Walters. 1983. Loss of duplicate gene expression in tetraploid *Chenopodium*. Biochem. Syst. Ecol. 11: 7–13.

Wit, F. 1958. Tetraploid Italian grass (*Lolium multiflorum* Lam.). Euphytica 7: 47–58.

Wohrmann, K., and H. Drew. 1959. Vergleichende Untersuchung über die CO_2—Aufnahme di- und tetraploider Pflanzen von *Trifolium incarnatum* im Abhangigkeit von Lichtenstät und Temperature. Züchter 29: 264–270.

Wolf, P.G., Soltis, D.E., and P.S. Soltis. 1990. Chloroplast-DNA and allozymic variation in diploid and autotetraploid *Heuchera grossulariifolia* (Saxifragaceae). *Am. J. Bot.* 77: 232–244.

Wolf, P.G., P.S. Soltis, and D.E. Soltis. 1989. Tetrasomic inheritance and chromosome pairing behaviour in naturally occurring autotetraploid *Heuchera grossulariifolia* (Saxifragaceae). Genome 32: 655–659.

Wong, S.C., I.R. Cowan, and G.D. Farquhar. 1979. Stomatal conductance correlates with photosynthetic capacity. Nature 282: 424–426.

Wright, S. 1938a. Size of population and breeding structure in relation to evolution. Science 87: 430–431.

Wright, S. 1938b. The distribution of gene frequencies in populations of polyploids. Proc. Natl. Acad. Sci. USA 24: 372–377.

Wright, S. 1940. Breeding structure of populations in relation to speciation. Am. Nat. 74: 232–248.

Wright, S. 1941. On the probability of fixation of reciprocal translocations. Am. Nat. 75: 513–522.

Wright, S. 1969. Evolution and Genetics of Populations. Vol. 2. Theory of Gene Frequencies. University of Chicago Press, Chicago.

Wullschleger, S.D., M.A. Sanderson, S.B. McLaughlin, D.P. Biradar, and A.L. Rayburn. 1996. Photosynthetic rates and ploidy levels among populations of switchgrass. Crop Sci. 36: 306–312.

Yang, T.W., and C.H. Lowe. 1968a. Chromosome variation in ecotypes of *Larrea divaricata* in the North American desert. Madroño 19: 161–164.

Yang, T.W., and C.H. Lowe. 1968b. A new chromosome race of *Larrea divaricata* in Arizona. West. Reserve Acad. Nat. Hist. Mus. 2: 1–4.

Yen, Y., and G. Kimber. 1990. Meiotic behaviour of induced autotetraploids in *Triticum*. Genome 33: 302–307.

Zadoo, S.N., R.N. Choubey, S.K. Gupta, and J. Sybenga. 1989. Meiotic chromosome association in diploid and tetraploid *Avena strigosa* Schrel. Genome 32: 972–977.

Zderkiewicz, T. 1971. Content of oil in different stages of ripe fruits of diploid and tetraploid cumin (*Carum carvi* L.). Acta Agrobot. 24: 121–127.

Zhang, D., and T. Sang. 1998. Chromosomal structural rearrangement of *Paeonia brownii* and *P. californica* revealed by fluorescence in situ hybridization. Genome 41: 848–853.

Zhang, D., and T. Sang. 1999. Physical mapping of ribosomal RNA genes in peonies (*Paeonia;* Paeoniaceae) by fluorescent *in situ* hybridization: implications for phylogeny and concerted evolution. Am. J. Bot. 86: 735–740.

Zhang, H., J.Z. Jia, M.D. Gale, and K.J. Devos. 1998. Relationship between the chromosomes of *Aegilops umbellulata* and wheat. Theor Appl. Genet. 96: 69–75.

Zhang, L., R. Pickering, and B. Murray. 1999. Direct measurement of recombination in interspecific hybrids between *Hordeum vulgare* and *H. bulbosum* using genomic *in situ* hybridization. Heredity 83: 304–309.

Zhao, X.-P., Y. Si, R.E. Hanson, C.F. Crane, H.J. Price, D.M. Stelly, J.F. Wendel, and A.H. Paterson. 1998. Dispersed repetitive DNA has spread to new genomes since polyploid formation in cotton. Genome Res. 8: 479–492.

Zhao, X.-P., R.A. Wing, and A.H. Paterson. 1995. Cloning and characterization of the majority of repetitive DNA in cotton (*Gossypium* L.). Genome 38: 1177–1188.

Zimmermann, K.F. 1968. Polyploidie bei Knalugras und Glatthafer (*Dactylis glomerata* und *Arrhenatherum elatius*. Dtsch. Akad. Landwirtschaftwiss. Berl. Tagungsber. 18: 68–72.

Zohary, D., and U. Nur. 1959. Natural triploids in the orchard grass, *Dactylis glomerata*, L., polyploid complex and their significance for gene flow from diploid to tetraploid levels. Evolution 13: 311–317.

Zuckerkandl, E., and W. Hennig. 1995. Tracking heterochromatin. Chromosoma 104: 75–83.

INDEX

Printed in the United States
By Bookmasters